Information Strategy and Warfare

This book discusses the decisive role that information and information strategy now plays in modern warfare. Information strategy is developed here as a new kind of power that can stand on its own as a distinct, increasingly influential tool of statecraft.

Three principal themes are explored in this volume. First, the rise of the "information domain" itself, and of information strategy as an equal partner alongside traditional military strategy. Second, the need to consider the organizational implications of information strategy. Third, the realm of what has been called "information operations." Throughout this book, information operations is portrayed as comprising all manner of conduits for information transmission, but also as being critically dependent upon good content.

The evolving importance of computer networks and the role of blogs in information strategy become apparent in the overall "battle for the story" that is being waged by all sides in the post-9/11 era. The need for good strategy is at the core of this book, and its deficit becomes painfully apparent.

This book will be of great interest to students of strategic studies, intelligence, terrorism and contemporary warfare.

John Arquilla is Professor of Defense Analysis at the Naval Postgraduate School in Monterey, California. He also serves as Director of the Information Operations Center. **Douglas A. Borer** is an Associate Professor in the Department of Defense Analysis at the Naval Postgraduate School in Monterey, California, and serves as a member of the Center on Terrorism and Irregular Warfare.

Contemporary security studies

NATO's Secret Armies
Operation Gladio and terrorism in
Western Europe
Daniele Ganser

**The US, NATO and Military
Burden-Sharing**
*Peter Kent Forster and
Stephen J. Cimbala*

**Russian Governance in the Twenty-
First Century**
Geo-strategy, geopolitics and new
governance
Irina Isakova

**The Foreign Office and Finland
1938–1940**
Diplomatic sideshow
Craig Gerrard

Rethinking the Nature of War
*Edited by Isabelle Duyvesteyn and
Jan Angstrom*

**Perception and Reality in the
Modern Yugoslav Conflict**
Myth, falsehood and deceit
1991–1995
Brendan O'Shea

**The Political Economy of
Peacebuilding in Post-Dayton Bosnia**
Tim Donais

The Distracted Eagle
The rift between America and Old
Europe
Peter H. Merkl

The Iraq War
European perspectives on politics,
strategy, and operations
*Edited by Jan Hallenberg and
Håkan Karlsson*

Strategic Contest
Weapons proliferation and war in the
Greater Middle East
Richard L. Russell

Propaganda, the Press and Conflict
The Gulf War and Kosovo
David R. Willcox

Missile Defence
International, regional and national
implications
*Edited by Bertel Heurlin and
Sten Rynning*

**Globalising Justice for Mass
Atrocities**
A revolution in accountability
Chandra Lekha Sriram

Information Strategy and Warfare

A guide to theory and practice

Edited by John Arquilla and Douglas A. Borer

Routledge
Taylor & Francis Group

NEW YORK AND LONDON

First published 2007
by Routledge
605 Third Avenue, New York, NY 10017

Simultaneously published in the UK
by Routledge
2 Park Square, Milton Park, Abingdon, Oxon, OX14 4RN

Routledge is an imprint of the Taylor & Francis Group, an informa business

© 2007 Selection and editorial matter, John Arquilla and Douglas A. Borer; individual chapters, the contributors

Typeset in Times by Wearset Ltd, Boldon, Tyne and Wear, UK

Library of Congress Cataloging in Publication Data
A catalog record for this book has been requested

British Library Cataloguing in Publication Data
A catalogue record for this book is available from the British Library

ISBN 13: 978-0-415-77124-5 (hbk)
ISBN 13: 978-0-415-54514-3 (pbk)
ISBN 13: 978-0-203-94563-6 (ebk)

In memory of Ted Sarbin

All royalties generated from sales of this book will be distributed by the Naval Postgraduate School Foundation to charities supporting service personnel killed and wounded in the War on Terror.

Contents

Tables

Notes on contributors

John Arquilla, PhD, is Professor in the Department of Defense Analysis at the Naval Postgraduate School Department of Defense Analysis in Monterey, California. His research interests include the revolution in military affairs, information-age conflict, and irregular warfare. He has published on these subjects as well as on special operations through history.

Frank J. Barrett, PhD, is Professor of Management in the Graduate School of Business and Public Policy at the Naval Postgraduate School in Monterey, California. He is also a Faculty Member of Human and Organizational Development at the Fielding Graduate University. Dr Barrett has written and lectured widely on social constructionism, appreciative inquiry, organizational change, jazz improvisation and organizational learning.

Douglas A. Borer, PhD, is Associate Professor at the Naval Postgraduate School Department of Defense Analysis in Monterey, California. His research interests include war and political legitimacy, economic statecraft, and strategy and systems, and has published on Vietnam, Afghanistan, and Iraq.

Dorothy Denning, PhD, is Professor of Defense Analysis at the Naval Postgraduate School in Monterey, California. Her areas of research include influence and information operations, cybercrime, terrorism, information security, and trust.

Maj. James Kinniburgh is an Air Force officer currently serving as Information Operations Branch Chief at the Joint Special Operations University, Hurlburt Field, Florida. Areas of special research include cybersociology, intelligence, influence and information operations, and special operations. He received his MS in Defense Analysis from the Naval Postgraduate School.

Carnes Lord, PhD, is Professor of Military and Naval Strategy in the Strategic Research Department of the Center for Naval Warfare Studies, US Naval War College. He is a political scientist with broad interests in international and strategic studies, national security organization and management, and political philosophy and has held various senior positions in the US government. His research interests include strategy leadership and statecraft, national security organization and management, and arms control.

Anthony R. Pratkanis, PhD, is currently Professor of Psychology at the University of California, Santa Cruz where he studies social influence and persuasion. Anthony Pratkanis is the founding and current editor of the scientific journal, *Social Influence.* He wrote this chapter while serving as a Visiting Professor of Information Sciences at the Naval Postgraduate School in Monterey, CA.

Glenn E. Robinson, PhD, is Associate Professor in the Department of Defense Analysis at the Naval Postgraduate School. He has traveled widely in the Middle East and has studied at a number of universities in the region. His research interests include the relationships between regional peace and domestic disorder in the Middle East, collective action in Muslim societies, and the political economy of authoritarianism and democratic transitions.

David Ronfeldt, PhD, is a senior political scientist at the Rand Corporation (he wrote for this volume while on leave, independently of any Rand project). His interests include information-age conflict, networks and netwars, and social evolution.

Hy S. Rothstein, PhD, is a senior lecturer in the Naval Postgraduate School Department of Defense Analysis. He served in the US Army as a Special Forces officer for 26 years before starting a teaching career. He spent many years in Latin America training and advising governments threatened by active insurgencies and lectured on related subjects. His interests include national security policy, unconventional warfare, combating terrorism, psychological warfare, and military deception.

Theodore R. Sarbin, PhD, was born in 1911 and passed away in 2005. He first served the Psychology Faculty at the University of California at Berkeley, from 1949–69. He proposed the unorthodox position that problems conventionally thought of as "mental illness" could better be construed as moral judgments rendered by those in a position of social power about individuals whose conduct is unwanted or perceived as dangerous. Professor Sarbin continued to challenge orthodox views in psychology throughout his professional life. He offered interpretations of hypnosis that avoided the necessity of positing a special mental state, viewing hypnotic behavior in terms of a person's ability to take the role of the hypnotic subject. Likewise, such concepts as "hallucinations," "anxiety" and "schizophrenia" were subject to Sarbin's relentless efforts to "demythologize" psychology. Professor Sarbin left Berkeley to join the faculty of the University of California at Santa Cruz in 1969. He continued there until his retirement in 1976. In addition, he served for varying periods on the faculty at the Naval Postgraduate School in Monterey. In 1987, he became a Research Psychologist for the Defense Personnel Security Research and Education Center (PERSEREC) a program of the US Navy. He continued to work at PERSEREC until June 2005, two months before his death at the age of 94. In the course of his academic career, Professor Sarbin received scores of honors – including both Fulbright and Guggenheim fellowships. He was a Research Scholar at Nuffield College of Oxford University in 1963. He was a Fellow at the Center for Advanced Studies at Wesleyan University in 1968–9, and returned there for another period in 1975. He received the Morton Prince Award from the Society for Clinical and Experimental Hypnosis and the Henry Murray Award from the American Psychological Association. He was recognized with a lifetime achievement award from the Western Psychological Association in 2001. Included among his more than 250 professional publications are six books and another six edited volumes. Over the past 20 years, Sarbin concentrated on developing and promoting the practice of narrative psychology, departing from the narrow research methodologies of traditional psychology in favor of a method based on the primacy of story and dramatic unfolding as a way of understanding human experience. Here again, Sarbin has been recognized as a pioneer in psychology, much as he was for his dissertation research done more than 65 years ago. This volume is dedicated to him.

Barton Whaley, PhD, is a Research Professor in the Department of Defense Analysis at the US Naval Postgraduate School and is a consultant with and author for the National Intelligence Council's Foreign Denial & Deception Committee (FDDC). An internationally renowned scholar and authority on deception and counterdeception, Dr Whaley was earlier associated with The American University's former Special Operations Research Office, M.I.T.'s Center for International Studies (1961–9), the RAND Corporation, and the Center for Strategic and International Studies.

Introduction

Thinking about information strategy

John Arquilla

Over 60 years ago, in the middle of World War II, Bernard Brodie explained the salience of sea power in the industrial era in *A Layman's Guide to Naval Strategy*.[1] Today, still quite early in the information age but already several years on in the first great war between nation-states and terrorist networks, it is imperative that the growing importance of information strategy be recognized and carefully studied. Much as Brodie noted that command of the sea allowed greater freedom of maneuver and access to the trade, wealth, and resources of the world while denying them to one's enemies, we intend in this book to outline the benefits that accrue to skillful practitioners of information strategy.

Exploring the strategic dimensions of the information domain is an undertaking much more complex than Brodie's. The elements of sea power have been well understood for centuries and have been codified in the classic studies of Alfred Thayer Mahan, Julian Corbett, and Raoul Castex, to name just a few, and the basic functions of sea power have changed little, despite enormous technological advances during the two centuries since Trafalgar. Navies are still interested in obtaining sea control by battle or intimidation, moving goods and troops wherever needed, and either bombarding or landing amphibious forces on a hostile shore – the principal challenge often being to try to do all these things simultaneously.

Information strategy, on the other hand, is a still-forming phenomenon that has both technological and nontechnological components and that encompasses both what one intends to do to the enemy and what one intends to do for oneself. The good information strategist must be the master of a whole host of skills: understanding the kind of knowledge that needs to be created; managing and properly distributing one's own information flows while disrupting the enemy's; crafting persuasive messages that shore up the will of one's own people and allies while demoralizing one's opponents; and, of course, deceiving the enemy at the right time, in the right way. Next to the nuances of the information domain, the demands of naval strategy pale.

In this volume, we address each of these aspects of information strategy. In doing so, we articulate three principal themes that surface and resurface in recurring arcs across each of the chapters. First, we emphasize and explore the rise of the information domain itself and information strategy's emergence as an equal partner alongside traditional military strategy. We have defined the informational realm far more broadly than just in terms of the traditional diplomacy practiced by state departments and ministries of foreign affairs. Our inquiry includes all of the persuasive efforts that directly target mass publics, and we examine in some detail the general management of information, data security, and the subtle play of psychological operations and deceptions.

The second major theme we develop is based on the notion that an undue focus on technology will leave one wandering in a labyrinth. This implies a serious need to consider the organizational implications of the emergence of information strategy. The network form, it will be seen, emerges as a central organizational component in the redesign of civil and military institutions. The fact that a wide range of terrorists, rogues, and transnational criminal enterprises have already gotten a head start in a new "organizational race" to form networks – akin to the Cold War-era arms race – should galvanize our efforts to focus on building networks of our own to fight these darker forms.

Our third theme concerns the broad, amorphous realm of what has been called "information operations" (IO), where some foundations of information strategy have already been laid. In our view, though, IO has been too narrowly depicted in the decade and a half since its modern reemergence as a major issue area for national security policy, and so our conception must be both broadened and deepened. Thus, throughout this study, IO is portrayed as both cross-cutting and inherently interdisciplinary in nature. It is seen as consisting of all manner of *conduits* for information transmission, but also as being critically dependent on valuable *content*. The very best wiring in the world simply cannot make up for poor wording.

Ultimately, the concepts of information strategy developed in this volume may even imply the rise of a new kind of power that can sometimes stand on its own as a distinct, increasingly influential tool of statecraft – one less dependent on the coercive power of the "big battalions" and much more reliant on the persuasive power of big ideas. Above, I mentioned the "reemergence" of information operations and associated strategies, which implies that much has come before and that the study of relevant historical events and trends can help guide the search for information strategic concepts in our time and in the years ahead. Even a cursory backward glance quickly reveals the value of such an exercise and calls us to be mindful that information strategy did not spring forth fully formed, like Athena from Zeus's head. It has formed and reformed, shifted shape and emphasis, for millennia. We ignore this long experience at our peril.

Some roots of information strategy

As is the case with most intellectual constructs, information strategy did not emerge without antecedents of some sort. The content-oriented aspects of information strategy, for example, have been around since ancient times. The biblical account of Gideon fooling the Midianites about the overall size of his field forces is a cautionary tale about the crippling effects of ignorance and the striking power of a "knowledge advantage." The early Romans, though, were far less interested in the various military applications of psychological operations and deception – they were pretty much straight-up fighters – and focused instead on the larger theme of crafting an image of fair, tolerant, and inclusionary rule. Thus, the prospect of living as a Roman citizen became the empire's "great attractor" to those on the outside. Later on, in the east, however, Byzantine information strategy found its way back to deception and also keyed on skillful management of information about threats to the frontiers. In this way, the almost always less numerous Byzantine forces were, for 1,000 years after the fall of Rome, nonetheless sufficient to meet the harsh demands of imperial defense. Smooth, swift flows of information were real force multipliers back then, much as they are today.

Through the end of the first millennium of the common era, practitioners tended to focus on or show greater strength in just one or another aspect of information strategy. But the Mongols of the thirteenth century were the first to demonstrate considerable prowess in virtually every dimension of this art. In terms of content-based information operations, for example, they quickly became aware of how both Muslims and Christians perceived the coming of the Mongols to be a form of divine punishment and a sign of the end of days. They exploited this, once they learned the term *tartarum* (denoting the nether world), by renaming themselves *tartars*, instilling the fear in one and all that they might not even be human. They hardly seemed to be.[2]

The Mongols were also masters in the use of psychological operations and deceptions as means of improving their chances of taking cities and conquering opposing armies. It required only a few examples of slaughter to encourage most cities to surrender to the Mongols' surprisingly efficient, often enlightened rule rather than stand a siege. And in open battle, while always on the offensive strategically, the Mongols' great mobility allowed them the luxury of feigning retreat – sometimes for days – in order to dislocate enemy forces prior to the climactic clash of arms.

The Mongols may also have been the best managers of information ever. Their Arrow Riders, a kind of medieval Pony Express, kept all throughout the empire informed of events and military developments with at most a week's lag – this over 700 years ago in an empire that covered much the same territory the Soviet Union did. In terms of command and control, the Riders were supplemented with an early kind of semaphore signaling system that allowed Mongol field armies to undertake such complex battle evolutions as the "Crow Swarm" and "Falling Stars" – both examples of omnidirectional attack doctrines that have resurfaced recently in American "swarm tactics."[3]

In short, the Mongols were masters of information strategy at every level, and they skillfully integrated it into their political and military calculations – a remarkable achievement for a nomadic, semiliterate people. But the lines of communication and trade infrastructure that they nurtured eventually contributed to their undoing, when their "connectivity" facilitated the swift spread of the Black Death throughout their empire as well as Europe, where about a third of the population died from plague between 1350 and 1400. Fear of debilitating computer viruses today is but a faint echo of the risks that come with connectivity, never better illustrated than in the case of the khans.

The decline of the Mongol Empire – still the largest empire the world has ever seen – left in its wake a kind of insularity of spirit when it came to the information domain. Instead of leading states embracing the diffusion of knowledge, sustained efforts were made in many fields to control its spread. Popes in Rome tried to prevent the printing of the Bible in the vernacular, despite the advent of movable type that made its mass production in multiple languages possible. Portugal's Prince Henry strove to control access to information about navigation, insisting that his prized charts be used only by Portuguese pilots. Even in commerce there was much evidence of efforts to protect information flows. For example, Venice kept its finest artisans at home, and if a few found their way abroad, the *doge* was not averse to kidnap or murder to stop the spread of knowledge about proprietary techniques of Venetian artisanship.[4]

But all these were just rearguard actions, even if they did go on for centuries. In the end, religious reformation could not be stopped, and the remarkable secrets of navigators and craftsmen also got out. The same was true in military affairs, where the marriage of gun and sail led to the ship of the line and where artillery, massed volley musketry, and other "best practices" transformed land warfare. All these innovations tended to be quite widely diffused. In the late eighteenth century, the Mahrattas of India were making the best artillery pieces in the world.[5] Still, the spread of such innovations tended to move at a relatively slow pace until the onset of the machine age that arose in the wake of Waterloo.

Industrial-age trends

Within a few decades of Napoleon's final defeat in 1815, the world was almost completely transformed by machines of all sorts – for manufacturing, transport, and, especially, information flows. The first giant leap in information technology since the printing press came with the electrical telegraph – there had previously been optical telegraphs that worked on line-of-sight relays of signals – which allowed information to move at a speed unheard of in all history. Shortly after the American Civil War, most of the world was linked by the telegraph, allowing information about events taking place virtually anywhere to be spread to everywhere else, within hours. Compared with the change in global sensibilities wrought by the telegraph, the coming of the Internet and the World Wide Web seems like a lesser phenomenon.[6]

In the realm of military affairs, the telegraph allowed for the command, control, and coordination of huge armies over vast areas. By the end of the Civil War, for example, Union commander Ulysses S. Grant was overseeing the operations of about a million troops distributed all over a battlespace roughly the size of Western Europe – with a core staff of twelve. As empowering as the telegraph could be, though, it created a new vulnerability: narrowly channelized information flows, the disruption of which would now likely have serious consequences. Thus, when William T. Sherman began operations aimed at capturing Atlanta in the summer of 1864, he had to hold back about 80,000 of his 180,000 troops to protect his rail and telegraph lines. This semicrippling drain on his forces was necessary because his communications were mortally threatened by a relative handful of Confederate raiders under the command of Nathan Bedford Forrest.[7] This was a foreshadowing of the information-age paradox that technological advances can simultaneously empower and imperil.

Just as important as its effects on military affairs, the telegraph created a kind of global social consciousness that helped catalyze mass movements opposed first to slavery and later to the excesses and brutality of colonial rule. In both instances, civil society movements were hugely energized by their connectivity. The success in reining in King Leopold's depredations in the Belgian Congo was a signal example of the growing power of social networks and a harbinger of human rights movements to come.[8]

At the turn of the twentieth century, there were also signs of the rise of what we think of today as peace movements. While the British were struggling with the Boer insurgency in South Africa and American forces were combating guerrillas in the Philippines, telegraphed reports of the often brutal fighting prompted hundreds of thousands of ordinary citizens in Britain and the United States to protest these wars. In the British case, this movement (which even featured a Cindy Sheehan-like figure – Emily Hobhouse, whom Lord Kitchener simply referred to as "that bloody woman") had a profound effect on the government's decision to offer fairly generous peace terms to the South African *Broeders*.[9]

Antiwar protesters in the United States got much less traction, an early example in a pattern suggesting that American mass publics do not falter in their support for a conflict simply because time drags on and casualties mount. This point has been borne out again and again in public opinion data gathered on the many American conflicts going back more than fifty years, ranging from Korea and Vietnam to the more recent invasion and occupation of Iraq and the war on terrorism. In each case, a clear pattern emerges: if a war becomes protracted, the ranks of those who think the intervention was a mistake grows, but significant numbers of Americans, usually a majority, nonetheless adopt a "stay the course" mentality. This was true even in the case of Vietnam, where, right up until the end of the war, about half of the American public wanted either to keep fighting as we were or to escalate military operations against North Vietnam.[10]

When World War I erupted a decade after the Boer War and the Philippine insurgency ended, a new kind of information strategy would come into its own: propaganda, or what we have more recently relabeled in a quest for a less pejorative term, "perception management," "public diplomacy," and "strategic communications." By whatever name, it is a process the goal of which is to influence mass publics by weakening the enemy's will, shoring up one's own, and persuading bystanders of the righteousness of one's cause.[11]

In the main, the Allies far outshone the Germans as information strategists during World War I, even in the technological realm, where they intruded on some of the Kaiser's secret communications with Mexico, which were aimed at inducing the latter to go to war with the United States. This "outing" of a most secret German diplomatic initiative had an electric effect on Americans, driving what had been until then a neutralist mass public to side increasingly with Britain, France, and the other Allies. And it was the American entry into the Great War that ultimately sealed the Germans' fate.[12]

In the propaganda realm, World War II followed the patterns of the First fairly closely. The Germans had improved their skills incrementally, particularly in the 1930s, when they successfully projected an intimidating image of military prowess in support of a string of bloodless diplomatic triumphs. But once the war began in earnest, the Allies largely had their way in portraying the Axis powers as evil incarnate.[13] In the area of technical information systems, the Allies also reprised their success from the previous war, this time to a much greater extent, as embryonic aspects of high-performance computing were employed to break the Axis powers' secret codes, leading to great Allied victories at Midway, Kursk, the protracted Battle of the Atlantic, and elsewhere.

Perhaps the single biggest difference between the two wars, though, in terms of information strategy, was the major impact of maturing radio and telephone technology on information management. For example, radio communications proved to be the lifeblood of the German blitzkrieg maneuver doctrine. On the defensive side, radio and telephone together enlivened the British Observer Corps and Fighter Command that defeated the Luftwaffe during the Battle of Britain. Even modern electronic warfare made its debut in World War II, with the jamming and counterjamming of airborne and ground-based radars featured in what Winston Churchill called "the wizard war."

The sensing, communications, and weapons guidance technologies introduced in World War II continued to evolve in the years that followed, driving the steady development of military information systems for the rest of the century. Even in the more ideational realm of the "war of ideas" that so defined the forty-year Soviet–American struggle between 1949 and 1989, radio remained a principal weapon in the arsenals of the information strategists on both sides. To be sure, television eventually began to provide compelling images from time to time, but it didn't truly come into its own, with advances like direct broadcast satellite, until the latter years of the Cold War, whose informational aspects were largely waged by what we call "old media." Even so, television did prove important on occasion, especially in places like Poland, where receivers sprouted up by the hundreds of thousands during the 1980s, helping fuel the deep discontent that eventually overthrew Soviet suzerainty.

What mattered even more than the conduits of information at this time, though, was the kind of *content* being conveyed by these various means. Here the basic message was summed up succinctly by Czech playwright and statesman Vaclav Havel: "Act as if you are free, and you soon will be." The grand master of this era, though, was Ronald Reagan, who – knowing that the Cold War simply could never be won militarily – conducted a relentless "war of ideas" during the 1980s, aimed at convincing the Soviets of the futility of their cause, and their subject peoples of the inevitability of liberation. Indeed, it was

Reagan who introduced the concept of "information strategy" into the policy lexicon (in his National Security Decision Directive 130).

The coming of the information age

In the waning years of the Cold War, the Internet was beginning to get on its feet, and the World Wide Web would soon follow. These many-to-many media made broad linkages possible among people around the world, holding together the ethnic diasporas of countless peoples and creating new kinds of "imagined communities" (to use Benedict Anderson's term) that were driven more by affinity (e.g. for environmental or human rights issues) than ethnicity. In short, a networked world was beginning to emerge, manifested in a form of "social globalization" mirroring the better known economic globalization phenomenon.

But right alongside this knitting-together process rose the shadow cast by those who understood that all the same information technologies that fostered peace, prosperity, and freedom could be used to foment social unrest and to disrupt the flows of commerce and strain the sinews of power and resource infrastructures. Again, all that empowered also imperiled, and the information age, it seemed, would be as fraught with peril as it would be full of hope for peaceful progress.

At the societal level, the double-edged nature of networking technologies was clearly grasped from the outset by Marshall McLuhan, whose *War and Peace in the Global Village* (1968; coauthored with Quentin Fiore) neatly captured the ambivalent temper of the times. In the military security realm, Thomas Rona, a technologist with the heart of a historian, observed the same problem emerging. His 1976 classic, *Information Warfare*, captured the essence of the problem that the same conduits that hugely empowered the advanced militaries would make them vulnerable to catastrophic disruption, much as the telegraph lines that had enabled the skillful control of large forces during the American Civil War had to be protected by large security detachments to avoid the dire consequences of their being cut.

In the decades since these two prophets of the information age first set down their thoughts, the dilemmas for society and security that have arisen have all followed the patterns they predicted. As McLuhan thought would happen, undue power has accrued to conduits of information, as opposed to the content being conveyed. Rona's insight was that a new form of strategic attack was emerging, akin to the rise of aerial bombardment, and much of the military thinking about IW – information warfare – with its emphasis on crippling enemy infrastructures, has borne him out. Indeed, the IW enthusiasts of our time sound much like the air power advocates of the 1920s and 1930s.[14]

Since McLuhan's and Rona's prescient writings, information strategy has been moving steadily along these two troubling tracks. The one devoted to the belief that medium trumps, or at least shapes, the message has led to an emphasis on using government-owned or controlled radio, television, and Web outlets instead of focusing on crafting messages that will be in demand by independent media outlets because of their thoughtfulness and persuasiveness.

The other developmental track of information strategy, the one that sees in the disruptive power of IW a new, largely bloodless, form of strategic attack, has led to a concentration of effort on technological aspects of infrastructural warfare in cyberspace, both offensive and defensive. This has led to a "hollowing out" of capabilities for and sensitivities to the nuances of psychological operations and deception and to a serious neglect of information management – a concept that is not even mentioned in the latest of the US Defense Department's official information operations roadmaps.

The war on terrorism

The first great conflict between nations and networks has been under way for over five years now, and information strategy has come strongly to the fore, in part because the US attempt to win the war outright by direct military action against suspect nations has done little to cripple terrorist networks – as reflected in the official statistics on significant terrorist incidents. In 2001, there were just over three dozen attacks worldwide. By 2004, there were over 3,000, according to the US National Counterterrorism Center.

Beyond the explosive growth in terrorist activity, another warning sign appeared in the wake of the US invasion of Iraq in 2003, when worldwide support for the US-led campaign against terrorists, rogues, and proliferators was gravely undermined. Throughout the Muslim world, the international community, and even in the United States itself, the consensus that had reigned since the 11 September 2001 attacks was fractured, imperiling coalition efforts in Iraq and on other fronts in the war on terrorism. Here was clear proof that the information domain could and sometimes would trump whatever might be going on in the military realm.

In addition to the divisiveness caused by the invasion of Iraq, revelations about abuses of detainees in Iraq, Guantánamo, and other sites put the United States in a deep hole in the "battle for the story." How could an American president talk about leading a global alliance against the modern barbarism of terrorism when his own minions were employing barbaric methods themselves? When the weight of numbers of deaths among innocent Iraqi civilians – variously estimated to be somewhere between 30,000 and 100,000 during the 2003–5 period – was added to other concerns about American behavior, support for the United States plummeted even further.

While all this was going on, al-Qaeda was still demonstrating some deftness in the realm of information strategy. At the highest level, its "grand narrative"[15] about the sacred mission to fight to reduce the shadow cast by American power on the Muslim world was still a powerful recruiting message. The continual stream of audio- and videotapes from Osama bin Laden and Ayman al-Zawahiri was proof positive not only that the terrorist network was still on its feet but also that it actually held the initiative in the information war – even the initiative to float peace overtures. And, at more tactical levels, al-Qaeda proved itself a skillful deceiver, increasing idle "chatter" from time to time in order to provoke the

Americans into expensive defensive evolutions (e.g. ratcheting up the "color code" warning level could cost as much as $1 billion a week).

Perhaps the most intellectually interesting strategic phenomenon of the war on terrorism is what it has demonstrated about the interplay of information strategy and military operations. The counterterrorism coalition's public diplomacy and other information operations have tended to be undermined by the amount of military activity going on simultaneously. Even some military successes seem to have had negative effects on the way the world perceives coalition efforts. By contrast, al-Qaeda and its affiliates and sympathizers have consistently used force far less often and far less effectively in a military sense, yet their reputation has seemingly been enhanced merely by their demonstrated ability to keep waging the war.

Thus, the war on terrorism seems an almost perfect laboratory for exploring the three themes that we believe resonate throughout the studies in the volume: the rise of the information domain and its interplay with military affairs; the growing power of networked organizations; and the multidimensional nature of information operations themselves. At this point, it seems that the counterterrorism coalition is still in need of much improved awareness and action in each of these areas, whether for purposes of bringing the al-Qaeda war to a successful conclusion or to prepare for the inevitable rise of new networks that will follow in the footsteps of bin Laden, the dark pioneer of an age of "netwar."

This volume is intended to offer some correctives to this troubling situation. We seek to revitalize earlier, much broader perspectives that are, in our view, more in line with the higher-level thinking that went on during the 1980s, when Reagan was winning the Cold War with an idea-driven information strategy. We also take more forward-looking perspectives on the rise and role of networks and networking and consider the full range of options available for conducting information operations and warfare against a range of adversaries, from terrorist networks to rogue nations.

In an era characterized by what the Pentagon calls "the Long War," when it is already apparent that the importance of information strategy is growing relative to that of military strategy, it seems clear that this is a struggle that will not be won simply by force of arms. In such a world, skillful information strategy is likely to prove the difference between victory and defeat.

An overview of the book

Information Strategy and Warfare is roughly divided into halves. In the first, the issues are addressed from a largely conceptual point of view. The role of ideas, the kind of knowledge we should seek, how influence works, and how our principal enemy thinks about the information domain – these are the principal topics that we believe form an essential basis for the development of a sound information strategy.

Frank Barrett and the late Ted Sarbin (to whose memory this book is dedicated) begin from the above-mentioned point about it being clear to all by now that, like the Cold War, the war on terrorism is unlikely to be won by military means. Although this realization has given a huge boost to the role of information strategy, it is not at all apparent, Barrett and Sarbin point out, that ideas can easily or happily be "sent off to war." Ideas, they suggest, are fundamentally persuasive rather than coercive, and hence the "war of ideas" metaphor must be used with great caution when thinking about the current conflict.

Indeed, Barrett and Sarbin muse about the possibility that the war metaphor should not be employed at all in the realm of ideas. In this regard, their basic argument connects in a cautionary way with our larger theme about the rising importance of the information domain itself. That is, they acknowledge the growing power of the ideational realm but caution that even thinking about ideas as "weapons" in this conflict may undermine their very potency. In a sense, Barrett and Sarbin are arguing that the rise of information strategy may be redefining war itself as a social phenomenon in which the use of force is increasingly becoming a backdrop to the more salient, more decisive public debates about the context and conduct of the conflict.

Knowledge is of course closely related to ideas, and in Chapter 2 David Ronfeldt considers what kinds of knowledge should be cultivated in the effort to wage a war on terrorism. His key concern is that over the past five years, far too much emphasis has been given to understanding the religious motivations of al-Qaeda and its affiliates. He advances the argument that al-Qaeda can be best understood as the first truly "global tribe." But it *is* still a tribe, with recognizable tribal organizational structures and practices. In particular, it turns out that the tribe is well suited to networking – much more so than the various hierarchies that have devoted themselves to al-Qaeda's destruction.

Ronfeldt's key insight for information strategy is that by focusing unduly on understanding and countering religious zealotry, we may have amassed and acted on the wrong kind of knowledge. To win this war, he argues, we must redirect our efforts to concentrate on understanding how globalism and tribalism have come together and mutated into a most virulent, predatory form. Making this intellectual-perceptual leap will require a major shift in investment in human capital. Doing so is not only feasible but absolutely necessary if we are to make real progress in the years ahead – against al-Qaeda and whatever new "global tribes" may rise in its wake.

In Chapter 3, Anthony Pratkanis evaluates the performance of American influence strategies undertaken at a number of levels, from psychological warfare by the US military against insurgents in Iraq to presidential-level efforts to influence the beliefs and preferences of the American public, the Muslim world, and the broader international community. The mixed results of these efforts over the past several years lead him to infer that our basic beliefs about the sources of influence over others may have to be reconsidered and further refined.

Pratkanis works from the perspective of an experimental social psychologist, making the case that a whole range of concepts commonly employed in commercial and social settings may also be applied to the war on terrorism. For example, the concepts of "commitment and consistency" can be used to wring initially small admissions and concessions that can be cultivated and grown into larger ones. "Social proof" could be useful in encouraging respected individuals in peer groups to become exemplars for others. Above all, influence may also be a function of "giving before getting" – that is, a willingness to launch some conciliatory initiatives may generate huge dividends, connecting with those who would then feel impelled to reciprocate any act of kindness. Seen in this light, Pratkanis's chapter constitutes an approach to revitalizing the psychological operations component of IO by tying these activities much more closely to proven practices employed with success in other areas of human interaction.

Glenn Robinson rounds out the first half of the volume by focusing systematically on understanding al-Qaeda's views about the importance of the information domain, as well as the terrorist network's development of what might be called a "network-centric" approach to information operations themselves. Robinson examines all three of the cross-cutting themes considered in *Information Strategy and Warfare*. He offers for the first time a serious net assessment of al-Qaeda's strengths and weaknesses when it comes to waging information warfare, both offensively and defensively.

In the main, Robinson finds that al-Qaeda is a competent practitioner of both information management and data security – witness its continued worldwide operations and our difficulties in breaking into or disrupting its financial flows. But al-Qaeda's greatest strength is undoubtedly made manifest as it wages its own "war of ideas" against the United States and its allies and integrates a small number of terrorist attacks with its information strategy. The best example of this was the 11 March 2004 attack in Madrid that helped topple a government aligned with US policy toward Iraq. Beyond simply assessing al-Qaeda's information strategy, though, Robinson draws from his analysis several new ideas about how to inform and guide our own strategic responses to this wily, determined, and very skillful foe.

In the second half of the book, we focus more specifically on prescriptive, policy-oriented ways to improve American information strategy. The topics covered range from public diplomacy to the dark arts of deception and psychological operations, and then on to a methodology for assessing others' capacities for waging the more technologically oriented modes of information warfare.

In Chapter 5, Carnes Lord reconsiders how American public diplomacy – the effort to tell our story to foreign mass publics – has been organized and conducted, beginning with a careful account of our previous initiatives in this domain. As a former public diplomacy policy maker, Lord casts the whole issue in a new, probing light. He recounts with some irony the steps that led to the dissolution of the US Information Agency (USIA) late in President Clinton's second term, and he assesses the mixed performance of American public diplomacy since the onset of the war on terrorism, with an eye toward making improvements.

For Lord, the great challenge now is to revitalize American public diplomacy by means of organizational redesign. Although he laments the USIA's having been folded into the State Department, his prescription for remedying the situation is to take a fundamentally networked approach, radiating outward from State and reaching into all the relevant nooks and crannies sprinkled about the bureaucracy. Indeed, this approach could well enliven American public diplomacy in ways that might not previously have been imagined. As Lord tells it, there may be an opportunity here to make a virtue of necessity. Let us hope so.

If public diplomacy is in serious need of fixing, then the situation as it applies to deception – one of the key components of information warfare – is absolutely grave. In Chapter 6, Barton Whaley, long recognized as one of the world's leading experts on deception, critiques the current state of affairs and recommends a clear-cut set of steps that can be taken to revitalize one of our most important but underutilized tools of information strategy. In Whaley's view, even our media-saturated, deception-phobic democratic society can still craft a highly effective capacity for conducting such tricky operations. And this can be done, he argues, at modest cost.

To support his argument for the cost-effectiveness of increased investment in deception, Whaley makes considerable reference to earlier cases in history in which deception capabilities have been successfully developed. The salient point is that the resources necessary for mounting even the most far-reaching and sustained deceptions are but a tiny fraction of those available. The key is to be willing to make a small investment – as little as 1 percent of military budget – but this willingness can exist only where there is a proper appreciation of the overall role of deception in the realm of information strategy. Even beyond the information domain, Whaley observes, a first-class capacity for deception will achieve greater economies in terms of reducing the amount of military force required to deal with a particular contingency. Restoring this neglected aspect of information strategy to a healthier condition may obviate the perceived need to approach every crisis by invoking the Powell doctrine's tired, increasingly counterproductive mantra of "overwhelming force."

Much as Whaley's chapter keys on the many difficulties associated with deception operations and the ways to cope with them, Hy Rothstein's study of the state of psychological operations (PSYOP) in Chapter 7 pursues a similar path. Like Whaley, Rothstein advances the argument that the current US approach to PSYOP is fundamentally flawed – both organizationally and doctrinally – and sets out a process by which matters may be remedied. Although he focuses in particular on issues related to the war on terrorism, the process he outlines should be seen as being applicable across the range of potential future crises and conflicts, including in major wars and in protracted great-power rivalries.

In this respect, Rothstein is really focusing on the interplay of the military and informational realms in the formation of balanced, well-integrated strategies. For him, the difference between victory and defeat lies in the overall quality of one's strategy. Even very solid information operations are unlikely to overcome a bad strategy – although Rothstein also notes that a bad information strategy can fatally undermine even a good military strategy. With considerations of this sort in mind, Rothstein shows the way toward a renaissance both in an important facet of information operations and, more generally, in the whole realm of strategic thought.

While much of information strategy revolves around persuasion, perceptions, and influence, the technological realm remains crucially important. The basic point about advanced information systems is that although they clearly help to make us more powerful and prosperous, our dependence on them also makes us more vulnerable. In Chapter 8, Dorothy Denning breaks new ground in developing a systematic approach to assessing others' capabilities for waging war against information systems. She then applies her framework to assessing two cases of potential threats to the stability of the international system: North Korea and Iran. Denning argues that the approach she has employed to help understand the IW capabilities of these two potential adversaries can be applied more generally to the assessment of various other countries' capacities for waging cyberwar.

At a policy level, Denning's framework should perhaps also be considered as a useful construct for IW and for conflict in general. For example, her framework for analysis should help us think about how the information domain itself is increasingly intruding on the military operational realm. Also, her method of assessing others' offensive capabilities should serve as a guide to those seeking both early warning intelligence and practical guidance for preparing defenses against attacks on information systems. Finally, of course, the "reverse of the medal" is that Denning's analysis can be used to think about how one's own offensives in cyberspace might be mounted. With regard to cyberspace, she and Jim Kinniburgh provide, in Chapter 9, some new insights into the rising importance of Weblogs. Indeed, their analysis suggests that, given the ability to share best practices quickly, setting loose the "blogs of war" may make profoundly good strategic sense.

Information Strategy and Warfare concludes with Doug Borer's summary of the key issues raised in the volume and his assessment of the implications of the various findings made, both for the further development of the field and for specific policy innovations in a number of areas. Borer ventures some thoughts about why information strategy is so difficult to master and whether it is truly growing in importance relative to more traditional military strategy. He also muses about the seeming omnipresence of networks in virtually every issue area and reflects on the growing need to reconsider what until now has been an undue emphasis on the technological aspects of information operations and warfare, in favor of a rekindling of older, more idea-based concepts like deception and psychological operations.

Borer's last thoughts go to the first ones raised in this book. He reminds us that identifying the right kind of knowledge needed for a particular conflict may be the single most important choice the information strategist makes. He also considers the possibility that a "war of ideas" is an oxymoron, that persuasion rather than coercion is the order of the day. If this is so, then the key may be to articulate the appropriate content, dispatch it by networked means, then hold the reins loose.

In the end, we hope readers will agree that information strategy offers a vision of the ultimate "effects-based" way to operate. Whereas forces on the battlefield can readily be commanded and controlled with ever-increasing precision – thanks to the rise of advanced information technologies – influence, deception, and persuasion campaigns will only be undermined by overcontrol. Indeed, the great challenge of information strategy may be that it demands of its practitioners a hitherto unimagined capacity for knowing when to wait and watch. An odd thought, perhaps, but one growing ever more apparent in the conflicts that bedevil these strange days.

Notes

1 See also his earlier work on technological change and naval affairs, *Sea Power in the Machine Age*, 2nd edn, Princeton, NJ: Princeton University Press, 1943.
2 This theme is nicely exposited in James Chambers, *The Devil's Horsemen: The Mongol Invasion of Europe*, New York: Atheneum, 1985.
3 Jack Weatherford, *Genghis Khan and the Making of the Modern World*, New York: Crown, 2004. On the Mongol horde's swarming capabilities, see especially p. 94.
4 Norbert Wiener chronicled some of the history of these efforts to control the diffusion of knowledge in *Cybernetics: The Human Use of Human Beings*, London: Oxford University Press, 1950.
5 See John A. Lynn, "Heart of the Sepoy: The Adoption and Adaptation of European Military Practice in South Asia, 1740–1805," in Emily O. Goldman and Leslie C. Eliason, eds, *The Diffusion of Military Technology and Ideas*, Stanford: Stanford University Press, 2003.
6 On the impact of the telegraph on politics and society in the nineteenth century, see Tom Standage, *The Victorian Internet: The Remarkable Story of the Telegraph and the 19th Century's On-line Pioneers*, New York: Walker and Company, 1998. Another relevant study is Daniel Headrick, *The Invisible Weapon: Telecommunications and International Politics, 1851–1945*, London: Oxford University Press, 1991.
7 An excellent analysis of Sherman's infrastructure vulnerabilities can be found in Albert Castel, *Decision in the West: The Atlanta Campaign of 1864*, Lawrence: University Press of Kansas, 1992, especially Sherman's own views about the fragility of his lines of communication – what he called "the delicate part of my game" (p. 277).
8 Adam Hochschild has studied the emergence of nineteenth-century civil society networks that opposed both colonial depredations and the slave trade. See his *King Leopold's Ghost*, Boston: Houghton Mifflin, 1998, on the effort to stop Belgian atrocities in the Congo and his *Bury the Chains* (Boston: Houghton Mifflin, 2005) for an account of nineteenth-century civil society campaigns against slavery.
9 Unlike Sheehan, Hobhouse was a spinster who had suffered no loss in her immediate family. Her outrage was driven primarily by learning of the deplorable conditions of Boer women and children in Kitchener's "concentration camps." Hobhouse and the "Ladies Committee" had a huge impact on public opinion, not just in Britain but throughout the world. On the final peace terms granted to the Boers, see Byron Farwell, *The Great Boer War*, London: Penguin Books, 1977, pp. 429–37.
10 For details on these data, see Eric Larson, *Casualties and Consensus: The Historical Role of Casualties in Domestic Support for US Military Operations*, Santa Monica, CA: RAND, 1996.
11 On this theme, see James R. Mock and Cedric Larson, *Words That Won the War: The Story of the Committee on Public Information, 1917–1919*, Princeton, NJ: Princeton University Press, 1939.

12 This story was best told in Barbara Tuchman's *The Zimmermann Telegram*, repr. edn, New York: Ballantine, 1985.
13 The eventually debilitating effects on German soldiers of the successful portrayal of their cause as unjust is one of the major themes in Richard Overy, *Why the Allies Won*, New York: W.W. Norton, 1996.
14 For an overview of official thinking about information warfare as a form of strategic attack, see Greg Rattray, *Strategic Warfare in Cyberspace*, Cambridge, MA: MIT Press, 2001.
15 Marc Sageman's term, introduced in his *Understanding Terror Networks*, Philadelphia: University of Pennsylvania Press, 2004.

1 The rhetoric of terror

"War" as misplaced metaphor

Frank J. Barrett and Theodore R. Sarbin

When President Bush referred to the 11 September 2001 terrorist attacks as the "first war of the twenty-first century," he was marking off the attacks as distinctive and perhaps unprecedented. This new "war" beckoned some to reach beyond common conventions to make sense of what was happening. For some the marking was literal: *The Economist* referred to 11 September as "the day the world changed." Others reach for similes or some semblance of familiarity ("It's like Pearl Harbor," or "It's like the day JFK was killed"). None of these depictions are satisfying. In fact, the attacks were so unusual and unprecedented, there was no ready framework or language to convey what had happened. The temptation is strong to grasp for a more familiar set of words, to draw on some preexisting framework to make sense of these disruptive events. What's interesting, for our purposes, is the search for metaphorical constructs, "as if" expressions, in an effort to understand what feels radically unfamiliar. Bush and his aides drew immediately on the accessible and facile language of war.

In this chapter, we explore the consequences of choosing the war metaphor. We begin by discussing the nature of metaphor and the tendency for metaphor to morph into mythology such that the discourse community that repeatedly uses a metaphor comes to interpret it literally. We then discuss the widespread use of the war metaphor as a framework for guiding action and implementing policy. Analyzing the "war on drugs," we argue that the warfare framework for guiding policy, while useful, also creates unintended consequences. We build on this foundation to make three points about the "war on terrorism" metaphor. First, we propose that the application of the war metaphor as a framework for understanding and responding to the terrorist attacks is an inappropriate choice. Use of the warfare lexicon to construe the struggle against suicide bombers has obfuscated meaning, constrained the repertoire of possible responses to the attacks, and produced several unintended consequences, some of them self-defeating. Second, we propose it would be more appropriate to cite the framework of criminality and justice to shape and implement policy related to terrorism. Third, we argue that by choosing the warfare metaphor, US policy makers have undermined their ability to study and understand the conditions that give rise to suicide bombers, terrorist cells, and others we have labeled as the "enemy," and as a result, US policies may exacerbate those very conditions. If policy makers were to approach the problem in the same manner that criminologists approach their tasks, their efforts are more likely to lead to useful insights for stopping terrorism. Finally, we discuss the often used phrase "winning the war of ideas" that guides intervention efforts, and we suggest that this language is self-limiting and fails to appreciate the dynamics by which people are likely to alter their beliefs.

From metaphor to myth

While metaphors aid in articulating challenging or difficult experiences, it would be mistaken to think of them only as figurative or ornamental devices. Metaphors posit a framework for selecting, naming, and framing characteristics of one domain by asserting a similarity with another, implying meanings that may not otherwise have been noticed. For example, in the metaphor "man is a wolf," the ravaging, predatory nature of man is given focus; by contrast, the simile "my love is like a red, red rose" focuses on the more delicate, beautiful nature of one's beloved blooming to fruition, going through seasonal changes, and so on. The domains of wolves and roses in these examples invite particular meanings. In this sense, metaphors are filters that suppress some details and emphasize others – in short, "organize our view of the world" (Barrett and Cooperrider 1990; Lackoff and Johnson 1980).

At first blush, sentences such as "my love is a rose" or "man is a wolf" violate one of the maxims of the cooperative principle, a set of rules that facilitate communication (Grice 1975). Speakers are expected to observe the maxim of quality that utterances should make sense, that speakers should utter true statements and avoid untrue statements. Since men are not literally wolves or flowers, the reader or listener experiences a momentary condition of uncertainty, of epistemic strain. Unless the context marks the utterance as metaphor, the reader or dialogue partner is in a problematic sense-making situation. The reader or dialogue partner can resolve the epistemic strain by engaging in one of three strategies. The first, the credo ("I believe") resolution, is to regard the statement as true. The second, the figural strategy, is to regard the statement as metaphor. The third, the nonsense strategy, is to regard the statement as absurd, a condition that would terminate the cooperative discourse. If the dialogical context does not signal that a metaphor is intended, the writer can make use of linguistic conventions such as quotation marks or can qualify the anomalous utterance – "men are flowers" or "men are wolves" – with a marker phrase, such as "figuratively speaking."

For most readers, the metaphor "man is a flower" is transparent. Hardly anyone would regard the utterance as the basis for a belief that men and flowers belong to the same taxonomic class. When we employ or hear these phrases in conversation, we recognize the "as if" quality of the descriptive terms; we would not employ the credo resolution (treating the figural expression as literal), which would support our engaging in such actions as recommending watering people so they grow like flowers.

We are concerned here with the transformation from metaphor to myth, occasions when metaphor is less transparent than these illustrative examples. Earlier studies (Chun and Sarbin 1970) support the inference that nontransparent, unmarked metaphors tend to pull for the credo resolution; they are more likely to be given a literal rather than a figurative interpretation. These submerged metaphors are candidates for reification, or what Ryle (1949) called a "category mistake," which occurs when "idioms applicable to one category of things [are] applied to another category of events." As we demonstrate in the next section, some metaphors, particularly those associated with warfare, are vulnerable to submersion and reification that lead to category mistakes.

The attraction of the war metaphor

The war metaphor lends itself to mythical transformation and is an attractive framework for understanding and implementing policies. Associations with the domain of combat and warfare have been insinuated into several other domains and influence daily life in ways that have become so habitual that we sometimes lose sight of the fact that these are "as if" constructs. Consider ways that we have borrowed terms from the domain of warfare and applied them to political activities and marketing: we create "think tanks," we "mobilize resources," "target a particular group" in order to "make an impact," or we say that someone "missed the target" in relation to achievements as a set of indicators. The military metaphor has been used to frame the way we think about medicine and health care. Diseases "invade" the body, "advance," and "attack the body"; antibiotics and our immune systems "fight" against infections, and so on. There are shortcomings and limitations of the "disease as warfare" metaphor. This way of thinking might lead to over-mobilization efforts when less active interventions are appropriate, an emphasis on short term goals and interventions over long term health, treating the body as if it were territory or a battlefield rather than a living person (see Board on Global Health, 2006).

Our usual referents for the term "war" are acquired early in life when we read history books that discuss the Revolutionary War, World War I, World War II, and so on. A few words about the domain of warfare: war is a conflict that occurs between nation-states; it is often associated with campaigns of accumulation and expansion or defense against invaders who wish to expand or accumulate territory or other valued property; war is a conflict between enemies who have fixed, non-negotiable positions such that going to war is a symbol that negotiations have ceased; wars have beginnings and endings, victors and vanquished; the enemy is regarded as evil, so extreme action is warranted; when one's country is at war, one feels compelled to adopt a stance of patriotic unity and support of one's nation.

Given the long history of warfare and writings about war, it is easy to see how warfare has come to carry a familiar set of meanings and how one might easily draw on the imagery and lexicon of warfare. Perhaps because warfare represents mobilization of activity within an extreme environment and against a defined enemy, the language lends itself to metaphorical transfer into a variety of other domains, especially as a way to rally support for some cause or endeavor. A politician who says, "We need to combat inflation" is drawing on a rhetorical device that posits inflation as an external enemy that needs to be defeated by a concerted, collective effort lest the nation be overtaken by the momentum of the inflationary forces. If inflation is construed as the enemy, elected officials become wary of supporting policies that might be inflationary, for fear of being seen as supporting the enemy. As we explore below, if we forget that these constructs are "as if" constructions and treat them literally, unintended consequences ensue.

The "war on drugs" is an example of metaphor that has achieved mythic status and influenced public policy implementation. A high priority of President George H. W. Bush, the war on drugs attracted public attention and catalyzed the implementation of strong legislation to criminalize drug trafficking. The framework through which actions were construed made it almost impossible to refer to drugs, drug traffickers, and drug users as anything but evil threats (see Edelman 1985 on symbolic acts of political leaders). References were made to scourges, criminals, cop killers, drug kingpins, and crack lords as enemies to be eliminated. Drug abuse was framed as an "assault" on American youth. Drug enforcement agents were referred to as "freedom fighters" helping to "take back our streets." By sheer repetition the metaphor and its entailments, the war framework has attained mythical status and has been used to justify extreme action, including US military intervention in Columbia, to curtail drug production.

While the metaphor was a useful framework for mobilizing commitment, there are ways in which it does not fit the reality and detracts attention from other factors. This is not a conflict between nation-states; the traffickers are not interested in capturing territory; there is no conceivable "peace" negotiation; and a victory is inconceivable – drugs will never be eliminated, so the continued existence of drugs can be interpreted as defeat. But worse, by framing antidrug efforts as a war on drugs and on drug traffickers, drug enforcers may not be addressing other important issues.

A recent RAND study demonstrated that the "war on drugs" campaign has framed the policy in such a way that resources are devoted primarily to interdiction and criminalization. The argument for focusing on enforcement and interdiction is that by going after traffickers, drug availability will decrease, prices will increase, and demand will decrease. The study authors pointed out that the policy does not distinguish between heavy and casual or recreational users. Arguably, the heavy users warrant more attention – they keep demand high and place a greater strain on the health care system. The RAND researchers found that for this population, drug treatment would be a more appropriate intervention. A policy that emphasizes warfare tends to construe a monolithic enemy, is indiscriminate, and misses such nuances.

The RAND study concluded that the war on drugs has had mixed results. Fewer resources and efforts are devoted to understanding the social and psychological conditions that make drug use attractive. Attention is deflected away from other approaches, such as prevention, treatment, legalization, and decriminalization, even though some evidence suggests that these methods are more successful than interdiction in reducing drug abuse (Nobles 2002).

Several other factors suggest poor outcomes in the war on drugs. First, since the war on drugs was initiated, the price of cocaine has declined, indicating greater availability on the streets and increased purity, making this drug more accessible and more damaging (Sterling 1999). Second, in no small part as a result of drug enforcement policies, the courts are clogged and the prison system is overcrowded. Finally, the policy fails to notice addiction to the two most widely used (and arguably the most damaging) drugs – alcohol and tobacco.

The war metaphor may be an inappropriate way to frame and address the drug problem. The war metaphor identifies forces of evil (as it must) and thus blinds us to important nuances; it emphasizes a mindset of "quick strikes" and aggressive interventions that diverts us from addressing the social factors that lead to drug use. Some have argued that the social problem of drug use is one that requires careful and patient analysis and pragmatic strategies that address an entire spectrum of interventions, prevention measures, and treatments. In the next section we discuss how the vocabulary of the "war on terrorism" lends itself to analysis along similar lines.

Mythical status of the "war on terrorism"

The "war on terrorism" (also often called the "war on terror") is a case of an opaque metaphor that has been uttered and repeated by government officials and media as a literal truth. The users of the warfare metaphor have forgotten the "as if" quality of the construction, and the metaphor has become myth: meaning closes down, speakers become highly committed to a point of view from which it is difficult to back down, disconfirming evidence is unwelcome, ignored, or suppressed, and resources are mobilized in support of the committed actions.

The transformation of the "war on terrorism" metaphor into a guide for action based on traditional meanings of the term "war" is revealed in the decisions President Bush made immediately after the terrorist attacks on the World Trade Center and the Pentagon. Hours after the attacks, (Woodward 2004) on Air Force One, Bush said to Vice President Cheney, "We're going to find out who did this, and we're going to kick their asses." At 3:30 p.m., at Offutt Air Force Base in Nebraska, in a video conference with the National Security Council staff, the president said that the terrorists had carried out "attacks on freedom, and we're going to define it as such." He created a war cabinet that deliberated and put together a response. When his advisers debated about whether to focus on the al-Qaeda network, some suggested a broader response. Finally Cheney said, "To the extent we define our task broadly, including those who support terrorism, then we get at states. And it's easier to find them than it is to find bin Laden." In President Bush's speech to the nation at 7 p.m. on 11 September, he expressed "disbelief, terrible sadness, and a quiet, unyielding anger." He encouraged the country to "stand together" and to "defend freedom and all that is good and just in our world." He imputed the motives of the attackers as evil: "Today our nation saw evil … we are targeted because we're the brightest beacon for freedom and opportunity." The attacks were "intended to frighten our nation into chaos and retreat." He promised retaliation and announced that we would make "no distinction between the terrorists who committed these acts and those who harbor them." The next day, he referred to the attacks as "acts of war" and said that "freedom and democracy are under attack" and that this is a "monumental struggle of good versus evil."

The president thus reached for the language of warfare immediately after the attacks. Other US national leaders picked up the language and joined in displays of patriotism and unity – Senate leaders singing the National Anthem in front of the Capitol, the public embrace between Democratic Senator Tom Daschle and President Bush. Newspaper syndicates and television networks exploited the war metaphor with daily news segments under the caption "The War on Terrorism." After 9/11, television viewers watched news reports with the phrase "Attack on America" subtitled as a byline. The president asked Congress for large increases in appropriations for the armed forces. Even today, nearly every speech made by government officials contains references to the war on terrorism, without recognizing or even suggesting that such a war has little in common with the wars between nations recorded in history books. The media reinforced the metaphor-to-myth transformation by reporting official briefings about the so-called war.

When using the war metaphor, the enemy is depicted as the embodiment of evil, and extreme action is justified to force the enemy into submission. Saddam Hussein was linked (without evidence) with al-Qaeda as a sponsor of evil forces. Once the war on terrorism was announced (it was never officially declared by Congress), it became thinkable to launch preemptive strikes; the framework of war provides a logic for expansive military action. Indeed, US foreign policy has been altered as a result so that preemptive action is required: it is now legitimate to go into any country that seems to have a connection to al-Qaeda. Additional powers were granted to the president, and democratic processes, including the involvement of Congress, were curtailed (Yoo 2005). Executive powers were expanded to give the president more authority in world affairs, and these expanded powers, in turn, were cited to justify "extraordinary rendition," a secret program to extradite suspected terrorists from one foreign state to another for purposes of "rough" interrogation (Mayer 2005). The same logic has been used to justify wiretapping, queries into private bank records, and the questionable detention of prisoners.

When one's nation goes to war, one is expected to sacrifice to support those who are fighting. The sacrifices are seen as visible signs that one is supporting the troops. It becomes unpatriotic to criticize the commander in chief. Proponents accused those who opposed an invasion of Iraq as weak willed or siding with the enemy. Karl Rove, President Bush's political strategist, said that after 9/11 the Democrats "saw the savagery and wanted to prepare indictments and offer therapy and understanding for our attackers" ("Democrats Demand Rove Apologize for 9/11 Remarks," *New York Times*, 4 June 2005). As the rhetoric of war is accepted as "fact," some seem to notice its political utility: administration acts that would otherwise be controversial are justifiable because we are "at war."

Of greater concern is the possibility that the war metaphor might be misleading. As mentioned above, warfare is associated with conflict between nation-states or groups seeking to become nation-states, and often it is associated with campaigns of territorial accumulation and expansion or defense against enemies who wish to expand or accumulate territory or other valued property. The current struggle against terrorism is not a conflict between nation-states. Neither party seeks to expand territory or take over territory. Al-Qaeda is a stateless band of jihadists who are globally distributed. The terrorist organization is not an exclusively hierarchical one in which orders flow through an established chain of command. Intelligence sources have made it clear that after apparently routing the Taliban forces in Afghanistan, US troops encountered no massed army of terrorists camping there, nor is it likely that they will. Making use of computer technology, the terrorists operate a far-flung network organization in which individuals and small cells can communicate directly with each other over the Internet without going through a hierarchical structure. A new metaphor, "netwar," was coined (Arquilla and Ronfeldt 2002) to distinguish the flexible political strategies available to networked organizations employing decentralized digital technologies from the rigid hierarchical strategies of traditional military organizations. Just as the Internet has challenged old structures of commerce and communication, it has rendered obsolete many standard military practices, at least where the enemy operates through a networked form of organization. Nevertheless, the Bush administration made efforts to link terrorist attacks with nation-states.

Wars are associated with clear endings, with victory for one party and defeat for the other. The war on terrorism does not have clear beginnings and endings; it is unlikely that the conflict will end with one party as victor and another as vanquished; there will be no peace treaties to mark an ending. A complicating factor in the use of the war metaphor is the military expedition in Afghanistan against the Taliban government. Such action might be classified as a literal war in that the forces of a coalition of nation-states engaged in warfare against the forces of another nation-state. The Taliban forces appear to have been defeated, and a new provisional government has been installed. Even with the defeat of the Taliban, the so-called war on terrorism has not ended. Senior government officials tell us that more terrorist attacks are inevitable and that no ending is in sight. In fact, some claim that the victory in Afghanistan was not complete, as Taliban forces have reemerged in various parts of the country. In May 2006, there was a riot in Kabul with hundreds of chanting protestors marching around the presidential palace, looting shops, and shouting "Death to America"; eight people were killed and 107 injured (*New York Times*, 28 May 2006). These are not events one would expect after a total victory by US forces.

The ambiguity of endings and definitions of victory can be seen in the Iraq conflict as well. In May 2003, President Bush attempted to symbolically and publicly signal victory when, live on national television, he landed aboard the

USS *Abraham Lincoln* to announce the end of hostilities in Iraq. As Bush congratulated the crew, a large banner on board the ship announced "Mission Accomplished." Today that symbolic marking of victory seems hollow indeed. Since Bush's announcement, there has been an increase in terrorist bombings, challenging the notion that the military hostilities have ended. Like the "war on drugs" the "war on terrorism" is unwinnable. If the goal of the terrorists is to create an emotion – fear or terror – it is difficult to determine when victory is achieved. It only takes a few major terrorist events, such as bombings in Madrid and Bali, to undermine any contention that terrorism is being defeated. Engaged in an ongoing war, the US citizenry does not feel a sense of peace.

If this is so, it would be a case not only of treating the opaque metaphor as a believed-in proposition, but also of the metaphor maker himself – having adopted the credo resolution – being used by the implications of the metaphor (Turbayne 1962). Shortly after the 9/11 attacks, the president referred to the assault as the first war of the twenty-first century, and the language of war quickly took command. Having publicly committed himself to the use of "war" in its literal sense, he is not likely to shift his position by explaining that "war" is only a figure of speech (in fact he has insisted on the literal sense; see, for example, Stevenson 2005). Applicable in this context is a distinction between the "master's metaphor" and the "pupil's metaphor" (Lewis 1939). The master (a mentor or any high-status person – in this case, the president) is free to choose from a number of metaphors; the pupil (in this case, the media or the public) is not free to choose. The pupil must construct meanings from the implications of the master's choice of metaphor. Different implications would have followed had the president chosen a descriptor from any number of less bellicose metaphors – for example, "program," "campaign," "agenda," "criminal prosecution" or "initiative."

The error in the use of the phrase "war on terrorism" is compounded when we examine the term "terrorism." A number of definitions have been proposed. Fromkin (1975) defined terrorism as "violence used in order to create fear; but it is aimed at creating fear in order that the fear, in turn, will lead somebody else – not the terrorist – to embark on some quite different program of action that will accomplish whatever it is that the terrorist really desires" (p. 683). Laqueur (2002) regards terrorism as "not an ideology or a political doctrine, but rather a method – the substate application of violence or the threat of violence to sow panic and bring about political change" (p. 71). Common to these and other definitions is the core meaning: terrorism is a political strategy the goal of which is to incite fear and uncertainty among the citizenry.

Notwithstanding its status as a fixture of current political discourse, the "war on terrorism" slogan contains a logical fallacy. To speak of waging a war against a political strategy is to speak in absurdities. A political strategy cannot be the object of a conventional war. To talk about waging war against terrorism is like advocating war on propaganda, or on diplomacy, or on other political strategies. Similarly, terrorism is not an ideology. Terrorism is an antagonistic set of practices, an instrumental means. If we focus on the abstraction "terrorism," we lose sight of the human actors who are moved to attack us, who consider it their duty to die in the service of their cause.

Thus far we have argued that the war metaphor is subject to mythical trans-formation and that in the case of the "war on terrorism" metaphor, the rhetoric has been misleading. In the next section, we further explore the consequences of adopting the war metaphor, namely, a failure to understand those who attack the United States.

An alternative framework: terrorists as criminals

We propose that a more constructive metaphor for framing the current conflict is the metaphor of criminology. Terrorist actions are crimes that require careful intelligence and patient police work, just as police investigators ferret out the participants in crime syndicates and mafia. In the tradition of the military, the goal of warfare is more likely to be "search and destroy" rather than "arrest and prosecute," which is the goal of the criminal justice system. Joseph S. Nye, former director of the National Intelligence Council, supports the development of the law-enforcement model. "Suppressing terrorism is very different from a military campaign. It requires continuous, patient, un-dramatic civilian work, including close cooperation with other countries. It will also require coordination within our government for a systematic approach that addresses what to do before, during, and after a potential terrorist attack" (2002, p. 205) In this connection, it is worth mentioning that it was not the military but intelli-gence programs conducted by government agents that aborted the plot by terror-ists to bomb New York's Lincoln and Holland tunnels in 1993 – and to destroy eleven American passenger planes in Asia in 1995 (see Benjamin and Simon 2002, 2005) These and other successes of the law enforcement community in derailing terrorist attacks are quickly forgotten in the face of a small number of high visibility failures.

The perspective of criminal theory would trigger a different set of questions and probably a different response to terrorist activities. Criminologists see crime as deviant behavior that violates social norms; the focus is on the deed and the antecedents that motivated the deed; actions are associated with damage or loss rather than the embodiment of an inherent evil force. Researchers study what perpetrators find attractive and seductive in accomplishing deviant actions; they seek to understand how a person "empowers the world to seduce him to crimi-nality" (Katz 1988); the steps in the process from humiliation to rage and from rage to sacrifice; the process of identity change that takes place, how a person's self-concept shifts in a way that motivates destructive behavior. There is in fact a large body of knowledge that sheds light here.[1] When criminal researchers dis-cover that for perpetrators some destructive actions have seductive possibility they also often suggest that perpetrators drift in and out of legitimate and illegiti-mate behavior, tend to create narratives that neutralize their destructive actions, deny responsibility of injury to the victims, tend to appeal to higher loyalties as motivators (see Sykes and Matza 1957). Researchers would be more likely to discover how criminal (terrorist) behavior is learned, how potential criminals interact with primary and secondary groups, how they develop contacts, drift into subcultures, learn techniques and skills. When criminal researchers discover that for perpetrators some destructive acts carry seductive possibility, they also often suggest that in many ways the perpetrators are not really that different from the rest of us. In short, if we were to approach terrorism as a crime, we would work harder to understand the terrorists, a theme we explore in the next section.

Seeking to understand the terrorists

One of the consequences of framing the struggle against terrorism as war and the depiction of terrorists as evil is that it might obfuscate efforts to understand the conditions that have given rise to terrorist activities. From available information, we can surmise that terrorists think of themselves not as members of an organized army representing a nation-state but as human actors with a mission, a commitment, and a willingness to die for their cause.

Criminologists know that it is imperative to understand the conditions and motivations that lead to crime. Soon after the 9/11 attacks, President Bush put forth his own interpretation of the motivations of al-Qaeda and Osama bin Laden. In a demonstration of anger and resolve during his address to a joint session of Congress and the American people on 20 September 2001, he said, "Americans are asking, 'Why do they hate us?' They hate what we see right here in this chamber – a democratically elected government. Their leaders are self-appointed. They hate our freedoms – our freedom of religion, our freedom of speech, our freedom to vote and assemble and disagree with each other." This has been the prevailing depiction by the Bush administration – the view that the enemy has no rational agenda but has a powerful desire to destroy the United States, inspired by hatred for our values and fueled by religious fervor. Although it is common to describe enemies in such totalistic terms, doing so does not aid in understanding what actually motivates the suicide bombers and other jihadists.

If we consult recent history, we learn that people who are identified as Americans are hated by many Muslims because America is the most recent Western nation to have stationed military troops in Islamic countries and thus symbolically polluted their lands. This recent justification for hatred reflects a remote history, told and retold in narrative form. The context for contemporary belief and action is provided by the mythology acquired through the cultural stories of massacre, pillage, looting, and raping by waves of European Crusaders in the eleventh and twelfth centuries. Crusader narratives, told and retold from the victims' standpoint, keep the mythology alive. The subtext of the stories contains messages of hatred. Reinforcing the mythology are the remnants of powerlessness, an effect of the colonial exploitation of the Middle East by Western nations in the nineteenth and twentieth centuries. This argument is not to gainsay the rise of militant extremism in response to prolonged humiliation by the West or to failed policies of the former colonial powers. The people of Islam, like us, live in a story-shaped world. The plots of the stories that are part of their birthright continue to be influenced by ancient Crusader narratives. The motivation for terrorist acts by contemporary Muslim actors arises from their assigning themselves a valued role in narratives of revenge (Ahmed 2002; Doran 2001).

Recent research on terrorists attempts to grasp the historical background of jihadism and the motivation of individual terrorists. Pedahzur's (2005) review of the records of suicide bombers reveals that revenge and retaliation for injustices are motivators. Many suicide bombers suffered the violent death of a family member or friend; most have suffered some form of humiliation; many come from groups and societies that feel marginalized and oppressed and have little sense of a viable or efficacious future. Within such a context, the bombers serve as heroes who inspire hope. Based on a twenty-year study, Pape (2005) argues that one common motivator of suicide terrorists is to liberate their homeland from occupation or foreign control. He argues that religion is not the prime motivator for acts of terror, but it provides a powerful moral driver that serves to boost commitment to take extreme measures to oust invaders. Some theorists speculate deeper motivations beyond political goals. Khosrokhavar (2005, p. 31) claims that the primary candidates for suicide bombings are young second-generation Pakistanis who have been marginalized, stigmatized, subjected to discrimination and scorn and have limited access to opportunities. They "feel that they are reestablishing their links with the Islamic societies from which they have been cut off. They also have the impression that, as actors, they are more important than the Western societies that stigmatize them believe them to be. In symbolic terms, this allows them to feel superior to the West that despises them" (Khosrokhavar 2005, 32).

These studies suggests that the suicide bombers are part of social groups that have strong ideological bonds and specific political goals, are loyal to one another, follow an ethos of heroism, and feel drawn to taking on an important role in a larger purpose. Each suicide bombing becomes a recruiting device that sustains a cult of martyrdom in which suicide bombers are honored as the courageous combatants in the vanguard of a heroic war. In short, the persons we identify as terrorists are men and women who are actors in a drama, participating in a narrative whose origins lie not in individual genes, mass psychopathology, or a penchant for evil, but in a historical process.

If we are to take the criminology framework seriously, it would make sense for the United States to learn more about the socializing process by which young Muslims become suicide bombers. Here the tools of sociology and psychology can provide useful insights. Moghaddam (2005) proposes a socializing process as a series of progressive steps, like a narrowing staircase. As individuals experience unfairness and relative deprivation and feel they cannot improve their situation or achieve greater justice, they are likely to move to the next step in the staircase – anger and frustration. At this point, they are open to influence by leaders who encourage them to displace their anger and frustration as aggression against an enemy. Those who remain angry and continue to perceive social injustice may become attracted to and engaged with terrorist organizations. As they learn about the goals and beliefs of the group, they come to see terrorism as a justifiable and noble action. As new recruits, they learn to follow the ways of previous recruits who have carried out bombings and are held up as martyrs and heroes for killing civilians. This avenue of action is likely compelling, and it is reinforced if the recruits continue to experience a lack of voice or a lack of procedural justice in countries of the Middle East. As they are recruited into terrorist cells, they are more likely to conceive the world in binary "us versus them" terms, a familiar process of in intergroup dynamics – a tendency to exaggerate differences between the in-group and the out-group. This would explain why they target US civilians and express little remorse on hearing reports of deaths of civilians.

Time magazine's three-hour interview (Ghosh 2005) with a twenty-year-old Sunni Muslim Iraqi who volunteered to be a suicide bomber supports this developmental model of socialization and sheds light on the incremental steps that lead to a willingness to die for jihad. "Marwan" (a pseudonym) is educated, economically secure and has studied the Koran. When his commander agreed to allow him to do a suicide mission, he called it "the happiest day of [his] life." Humiliation was indeed a motivator for Marwan. He was inspired to become a suicide bomber after he witnessed an injustice: he said that in April 2003 he saw US soldiers fire on a crowd of demonstrators at a school, killing twelve and wounding more. He sought revenge, and eventually he was linked with Islamic radicals and was socialized into jihadist ways: "I read about the history of jihad, about great martyrs who have gone before me. These things strengthen my will." The socialization is extensive. Leaders of the organization show recruits videos of successful suicide hits and sometimes encourage them to visit sites of previous bombings for inspiration. Volunteers go through a self-purification ritual similar to the one Muslims undergo before embarking on a pilgrimage to Mecca. His overriding goal, Marwan said, was to force American troops out of Iraq. He also said there was no shortage of volunteers.

What is important to glean from these early studies is that applying the labels "fanatics," "axis of evil," and "freedom killers" to jihadists might obfuscate the complex reality: suicide bombers are rational, choiceful actors who are socialized into groups that offer credible and appealing revenge narratives. This raises a fundamental question: Does going to "war" against terrorism stamp out the perceived legitimacy of these narratives and make them less attractive?

Some contend that the war on Iraq has emboldened terrorists and inspired new forms of terrorist attacks. The Madrid train bombings were not carried out by al-Qaeda operatives. The cell responsible for them began to plan the attack the day after they heard a tape of Osama bin Laden condemning Spain's support of America in the war in Iraq. (See Benjamin and Simon 2005). Just as the metaphor "war on drugs" distracted us from rooting out the causes of drug use, the metaphor of a "war on terrorism" may be distracting us from seeking to understand and reduce the causes of terrorism, a point we explore below.

Invading Iraq may have increased the conditions – hopelessness, deprivation, sense of injustice, and so on – that spawn narratives that justify and encourage a cult of heroism for suicide bombers.

Terrorists target Americans for a variety of reasons: anger, resentment, frustration, poverty, and fear. Whatever the motivation, the terrorists have committed serious crimes and should be brought to justice. Their actions should not be interpreted as rightful revenge against Americanization, retaliation against injustice, or the legitimate struggle of freedom fighters. However, by framing our actions as acts of war, we may render the terrorists' motivations as understandable or even justifiable within the contexts of responses to war.

It should go without saying that attempting to understand the motivations of terrorists is not to condone their actions. Such understandings may provide the basis for alternative strategies for dealing with the conditions that influence people in the Islamic world and elsewhere to engage in terrorist actions. Framing the terrorist agenda as war makes it difficult to understand why some in the world are motivated to hate the United States and consider it an honor to die for their cause; it becomes difficult to grasp the narratives and socializing processes that attract young Muslims to become jihadists.

The "war of ideas"

In the previous section we proposed that young Muslims become terrorists for a variety of reasons: they believe the United States is a superpower that is "Americanizing" the Islam culture, they see bombings and invasions as an extension of the Crusades, they live in a culture of deprivation and fear, and they see injustice in US support of Israel's treatment of Palestinians. Each of these beliefs is reinforced by narratives in which events and details can be cited in support of America as the Great Satan, a strong recruiting device for terrorist cells. The terrorists understand the power of constructing credible narratives that includes images of heroes and martyrs dying for a cause. Bin Laden openly compares the presence of American troops in Muslim lands as a continuation of twelfth-century crusades. Where the terrorist organizations are successful, they are able to offer recruits an antidote to a sense of deprivation, membership in a group of like-minded people, coherent interpretations and meaning, models of others who have transcended deprivation by offering themselves as sacrifice, and narratives of a hopeful future.

The United States is also attempting to place terrorist activities within a larger narrative: it portrays suicide bombers as destroyers of civilization and wanton murderers of innocent civilians, and American troops as freedom fighters liberating Iraq. It is in this context that US leaders speak of the need to win the "war of ideas" in the Middle East. Here again we run up against the limitations of an "as if" metaphorical construct. The metaphor of war calls forth images of attack and defense. This is an unfortunate frame for understanding the relationship between conflicting ideas and differences in beliefs. A country cannot send ideas off to war to claim the territory of the mind or to reclaim the mind-space now held by radical jihadists. When US strategists say that the mission is to "bring liberty and democracy to the region" or to "push democracy" as a "forward strategy," they are speaking of ideas as if they are weapons. Ideas, beliefs, values, attitudes, and theories are abstractions that do not march or attack.

Framing efforts as a war of ideas legitimates the notion that if ideas are delivered in a powerful way, they will "win the hearts and minds" of the Iraqi people. The State Department hired advertising executive Charlotte Beers, famous for her branding campaigns in global marketing, to improve the American image in the Middle East. She helped create pamphlets and television ads that depicted religious tolerance and Muslims living peacefully in the United States. (Several Arab countries refused to air the ads.) The notion that beliefs and ideas can be "delivered" may be behind the revelation that pro-American news articles written by US troops were placed in Iraqi newspapers and that Iraqi reporters friendly to America were paid to write articles portraying America in a positive light. The army employs the Lincoln Group, a Washington-based public relations firm, to translate articles into Arabic (*New York Times*, 3 December 2005).

Which narrative will win the day – America as the great Satan or America as the world liberator? In terms of daily existence, it must be hard for Iraqis to discern the benefits of the US invasion and Saddam Hussein's overthrow. Infrastructure is depleted; electricity, water, and sewage operations have not been restored to prewar levels. The loss of Iraqi lives when compared with the loss of American lives no doubt exacerbates resentment. Iraqis do not feel safe and secure. According to *The Economist* (29 December 2004), "There is only one traffic law in Ramadi these days: when Americans approach, Iraqis scatter. Horns blaring, brakes screaming, the midday traffic skids to the side of the road as a line of Humvee jeeps ferrying American marines rolls the wrong way up the main street." A documentary that shows Muslims peacefully practicing their religion in the United States is not likely to "win hearts and minds" under these circumstances. The claim is made that every bombing raid carried out by Western nations in which innocent civilians are killed or maimed increases exponentially the number of potential terrorists among the families, clans, and tribes of the victims. Using air attacks to root out small bands of insurgents on the ground does not help create a narrative of America as the just liberator. Campaigns of "shock and awe" and widespread destruction by American bombs invites storytelling and meaning-making that frame American efforts as belligerent rather than as an instrument of furthering universal justice. Such destruction makes it easier to depict terrorists as freedom fighters struggling against injustice or occupation or for self-determination. Bombing the enemy into submission is not likely to win them over. In the absence of hope for a better life, revenge narratives, reinforced by the promise of a privileged afterlife, become compelling. Delivering one-way messages to the Islamic populace is not likely to discourage young Muslims from turning to radical actions or to induce them to embrace universal values of democracy, especially when there are several other cues in play that reinforce a different narrative.

The larger question is not which message is more convincing but whether the notion that "war of ideas" is a useful way to depict different interpretations. Studies in communication research suggest that successful persuasion as a two-way process in which sender and receiver operate within the contexts of their respective frames of reference and make those frames known to one another. Communication is persuasive when there is a reciprocal process of exchanging information and developing shared meaning; credibility is achieved when both parties spend as much time on negotiation and learning as they do on delivering the solution, outcome, or "preferred belief." (This is not conceivable when we think of ideas as attacking and defending.) Communication theorists propose a model of "invitational rhetoric" as a way to persuade those whose values and worldviews are different from one's own (Foss 2004). The assumptions behind a model of invitational rhetoric are that different perspectives are valuable resources, that change happens when people choose to change themselves, and that all participants are open to being changed by the interaction. In order to convince others, one must look convincible.

Holding on to the notion that ideas can invade and intercede, detracts attention from the complex dynamics in the Middle East – the cultural and economic dynamics, the anguish, the social tragedy, the ethnic tensions, the tribal loyalties, the corruption, the identity conflicts, and the deprivation and humiliation that some think contribute to the rise to radicalism. The US vision of a democratic Iraq – one with elections in which multiple parties are represented and people are free from fear and intimidation – fails to appreciate the historical and cultural context of the Middle East. A bias for noticing the "hunger for freedom" in Iraq might blind leaders from noticing the other hungers that compete for fulfillment and attention.

The logic provided by the Bush administration – that war is justified because terrorists "hate freedom" – may not be convincing for long. By focusing on sending ideas off to war, we forget that war activities are a human responsibility, a series of actions and decisions that invite meaning-making efforts and narrative constructions that are beyond the control of the message senders.

Conclusion

Metaphors are powerful devices for sense making: metaphorical constructs allow us to place the unfamiliar within a familiar framework and disparate fragments of experience can be linked meaningfully. We naturally draw upon metaphorical constructs so as not to live in a world of confusion, equivocality, and paralysis. It's important to remember that no metaphor is true or false, that metaphorical frameworks cannot be agreed upon on the basis of some objective truth criteria. The metaphor "war on terrorism," is useful for a number of reasons: it helps to make sense of an unprecedented conflict, has emotional appeal, is useful for mobilizing support, inspiring patriotism, creating a sense of unity. The war metaphor spawns entailments and causal attributions (including labeling the enemy as "fanatics" or "axis of evil" or "freedom killers") that create a sense of familiarity that allows one to navigate actions.

It's also important to acknowledge that metaphors simplify complexity; if we are not aware of how metaphorical frames are delimiting our actions, we run the danger of over-simplifying the world and forgetting that we are operating within sense-making constructions. In other words, we run the risk of becoming prisoners of our own language. Just as the "war on drugs" metaphor has become habitual, the users of the "war on terrorism," metaphor forget the "as if" quality of their constructions that guide beliefs even in the absence of empirical evidence that would support such beliefs. Once this essential element is forgotten, the users become committed to a point of view that legitimizes one set of actions that might not be appropriate to creatively address a problem or issue.

In this chapter we have argued that a metaphorical framework that triggers labels and actions such as "quick strikes," "search and destroy," may not be appropriate for eliminating terrorism. In the long term, the Western nations should implement political, economic and psychological initiatives aimed at dissipation of the energy of those long-held Crusader narratives that encourage hatred and hopelessness. An alternative set of narratives should be created that would feature coexistence, tolerance, and mutual respect. The narratives would flow from economic and political acts initiated and supported by the Western nations. A program like the post-World War II Marshall Plan would provide economic benefits, which in turn would be the source for stories that could engender hope. Handing over large sums of capital to corrupt or tyrannical leaders, however, would not serve the purposes of substituting benign for malevolent narratives. Controls would have to be in place to ensure that economic and political benefits would accrue directly to impoverished citizens.

Student exchange programs could help break down the barriers to appreciating the values of distant cultures. Western leaders should foster engagement and dialogue with the leaders in Islamic nations, especially those who reject the tactics of violence of the fundamentalist wing of Islam. Whatever programs are implemented, to be effective they should have the imprimatur of the United Nations and other international organizations dedicated to security in a fractured world.

Changing the narrative worlds of a people dominated by hatred and hopelessness is a monumental task, but it is one that is potentially more productive for the world's security than the fomenting war abroad and organizing an unwieldly hierarchical bureaucracy under the banner of homeland security.

Note

1 Katz' theory (1988) construes all forms of criminality as moral responses to humiliation, a theme that resonates with the formation and socialization of terrorists in the Islamic world.

Acknowledgement

The authors wish to thank friends and colleagues for advice and insightful conversations that contributed to the development of the ideas in this paper: Barnett Pearce, Rich Gula, Dale Coke, Dan Duggan, Mark Nissen, John Arquilla, Madelene Coke Barrett, Ralph Carney.

References

Ahmed, A. S. (2002) *Islam Today*, London: I. B. Tauris and Co. Ltd.

Arquilla, J. and Ronfeldt, D. (2002) *Networks and Netwars*, Santa Monica, CA: RAND Corp.

Barrett, F. J. and Cooperrider, D. (1990) "Generative Metaphor Intervention: A New Approach to Inter-group Conflict," *Journal of Applied Behavioral Science*, 26: 223–44.

Benjamin, D. and Simon, S. (2002) *The Age of Sacred Terror*, New York: Random House.

Benjamin, D. and Simon, S. (2005) *The Next Attack*, New York: Holt and Company.

Board on Global Health (2006) Ending the War Metaphor: The Changing Agenda for Unraveling the Host-Microbe Relationship – Workshop Summary. Washington: National Academies Press.

Chun, K. and Sarbin, T. R. (1970) "An empirical study of 'metaphor to myth transformation'," *Philosophical Psychology*, 4: 16–20.

"Democrats Demand Rove Apologize for 9/11 Remarks," *New York Times*, 4 June, 2005.

Doran, R. S. (2001) "Somebody Else's Civil War: Ideology, Rage, and the Assault on America," in Hoge, J. F. Jr. and Rose, G. *How Did This Happen? Terrorism and the New War*, New York: Public Affairs, pp. 31–52.

Edelman, M. (1985) *Symbolic Uses of Politics*, second edition, Urbana: University of Illinois Press.

Foss, S. (2004) *Rhetorical Criticism: Exploration and Practice*, Prospect Hts, Ill: Waveland Press.

Fromkin, D. (1975) "Strategy of Terrorism," *Foreign Affairs*, 53, 683–98.

Ghosh, A. (2005) "Inside the Mind of an Iraqi Suicide Bomber," *Time*, 26 June.

Grice, H. P. (1975) "Logic and conversation," in Coe, P. and Morgan, L. (eds) *Syntax and Semantics (vol. 3) Speech Acts*, New York: Academic Press, pp. 305–31.

Katz, J. (1988) *Seductions of Crime: Moral and Sensual Attractions in Doing Evil*, New York: Basic.

Khosrokavar, F. (2005) *Suicide Bombers: Allah's New Martyrs*, London: Pluto Press.

Lakoff, G. and Johnson, M. (1980) *Metaphors We Live By*, Chicago: University of Chicago Press.

Laqueur, W. (2002) "Left, right, and beyond," in Hoge, J. F. Jr. and Rose. G. *How did this Happen: Terrorism and the New War*, New York: Public Affairs, pp. 71–82.

Lewis, C. S. (1939) *Rehabilitation and Other Essays*, London: Oxford University Press.

Mayer, J. (2005) "Outsourcing Torture: The Secret History of America's 'Extraordinary Rendition' Program," *New Yorker*, 14 February.

Nobles, D. (2002) The War on Drugs: Metaphor and Public Policy Implementation. Doctoral Dissertation, The Fielding Graduate Institute. Santa Barbara, California.

Nye, J. S. (2002) "Government's Challenge," in Hoge, J. F. Jr. and Rose, G. *How did this Happen: Terrorism and the New War*, New York: Public Affairs, pp. 199–210.

Pape, R. (2005) *Dying to Win: The Strategic Logic of Suicide Terrorism*, Random House.

Pedahzur, A. (2005) *Suicide Terrorism*, Cambridge: Polity Press.

Ryle, G. (1949) *The Concept of Mind*, Chicago: University of Chicago Press.

Sterling, E. (2004) "Drug Policy: A Challenge of Values," in Juday, E. and Bryant, M. (eds) *Criminal Justice: Retribution and Restoration*, Haworth Press, pp. 51–81.

Stevenson, R. W. (2005) "President Makes It Clear: Phrase Is 'War on Terror,'" *New York Times*, 4 August.

Sykes, Gresham and Matza, David. (1957) "Techniques of Neutralization: A Theory of Delinquency," *American Sociological Review*, 22(6): 664–70.

Turbayne, C. (1962) *The Myth of Metaphor*, New Haven: Yale University Press.

Woodward, R. (2004) *Plan of Attack*, New York: Simon and Schuster.

Yoo, J. (2005) *The Powers of War and Peace*, Chicago: University of Chicago Press.

2 Al-Qaeda and its affiliates

A global tribe waging segmental warfare

David Ronfeldt

As coeditor John Arquilla points out in the Introduction, "identifying the right kind of knowledge needed for a particular conflict may be the single most important choice the information strategist makes." In this chapter, I address this important insight by examining how al-Qaeda is viewed by US analysts and strategists.

It has become sensibly fashionable to regard al-Qaeda as a cutting-edge, postmodern phenomenon – an information-age network, or network of networks. But this emphasis misses a crucial point: Al-Qaeda and affiliates are using the information age to reiterate ancient patterns of tribalism on a global scale. They are operating much like a global tribe waging segmental warfare.

My purpose in this chapter is to describe the dynamics of classic tribes – what motivates them, how they organize, how they fight – and show that al-Qaeda fits the tribal paradigm quite well; I argue that the war they are waging is more about virulent tribalism than about religion. The tribal paradigm should be added to the network and other modern paradigms to help formulate the best policies, strategies, and analytical methods for countering it. What follows is a slightly revised edition of an earlier paper (Ronfeldt 2005) and an op-ed piece (Ronfeldt 2006). I have added a postscript to elaborate on the earlier discussion.

Trends in analysis

According to the latest thinking, al-Qaeda is now more important as an ideology than an organization, as a network than a hierarchy, and as a movement than a group. It is increasingly amorphous, although initially it seemed tightly formed. Osama bin Laden's core group may even be too weakened to matter very much in orchestrating specific operations.

These changes represent a considerable evolution for al-Qaeda as well as for expert thinking about it. Initially – before and after the 11 September 2001 attacks in New York and Washington – analysts wondered whether this mysterious organization was structured like a corporation, a venture capital firm, a franchise operation, a foundation, a social or organizational network – or all of these. Today, now that al-Qaeda has more affiliates, the network and franchise concepts remain in play, but the emphasis is on al-Qaeda's evolution into a decentralized, amorphous ideological movement for global jihad.

Since so little about al-Qaeda's organization is fixed, counterterrorism analysts and strategists have to be ready to adapt their views to shifting realities and prospects. For example, a major new strike on American soil directed by bin Laden might jar analysts back to a belief that al-Qaeda's core remains (or has recovered as) a strong, central unit with an effective capacity for command and control. Also, while al-Qaeda may look amorphous, the deeper reality may be that it is polymorphous, deliberately shifting its shape and style to suit changing circumstances, including the addition of new, semiautonomous affiliates to the broader network. And that raises a further reason for analysts to remain flexible: Clear as it may be that al-Qaeda and its affiliates are organized as a network, evidence is still lacking about many design details.

It is not enough to say that something is a network. According to one model (Krebs and Holley 2002), a network may start out as a set of scattered, barely connected clusters, then grow interconnections to form a single hub-and-spoke design, then become more complex and disperse into a multihub "small world" network, and finally to grow so extensive, inclusive, and sprawling as to become a complex core/periphery network. For a while, the pressures put on the al-Qaeda network evidently reduced its structure from a hub-and-spoke design back to a scattered-cluster design. But now it is growing again, apparently into a multihub design. Which design is it? Do the pieces consist of chain, hub (i.e. star), or all-channel subnets? And where are the bridges and holes that may connect to outside actors? The answers matter, for each design has different strengths, weaknesses, and implications. Some designs may be vulnerable to leadership targeting, others not. As research proceeds on how best to disrupt, destabilize, and dismantle networks, analysts are finding that in some cases it may be best to focus on key nodes and in other cases on key links; in some cases on middling rather than central nodes or links; and in other cases on peripheral nodes or links. But this view remains tentative. Moreover, much less is known about how to analyze the capacity of networks to recover and reassemble after a disruption (including the possibility of their morphing into a different design).

In short, analysts and strategists have adopted a basic set of organizational views to work with, but they still face a lack of knowledge about al-Qaeda and its affiliates, particularly about how they may combine and shift among network, franchise, hierarchical, and possibly other design elements. Thus, it is advisable for analysts not to become fixed on any one view but instead to work with "multiple models" (Davis and Arquilla 1991) whose content and probability may continue to vary. It is also advisable to keep looking for additional views that have not yet been fully articulated.

Here is a viewpoint worth adding to the mix: The organization and behavior of al-Qaeda and its far-flung affiliates are much like those of a classic tribe, one that wages segmental warfare. This view overlaps with the network view but has its own implications. It shows that al-Qaeda's vaunted, violent fundamentalism is more a tribal than a religious phenomenon. It also shows that continuing to view al-Qaeda mainly as a cutting-edge, postmodern phenomenon of the information age misses a crucial point: Al-Qaeda is using the information age to revitalize and project ancient patterns of tribalism on a global scale.

My main purpose in this chapter is to urge thinking more deeply about the tribal paradigm and its applicability to al-Qaeda. The tribal paradigm may have useful implications for US policy and strategy – especially for conducting the ideological "war of ideas" – although these are given only some preliminary attention at the end of the chapter.

Basic dynamics of classic tribes

As people banded together to constitute primitive societies thousands of years ago, the first major form of organization to emerge was the tribe. Its key organizing principle was kinship, as expressed through nuclear and extended family ties, lineage segments (notably clans) that spanned various families and villages, and claims of descent from a common, often mythologized, even godlike ancestor. The tribe's key purpose or function was to infuse a distinct sense of social identity and belonging, thereby strengthening a people's ability to bond and survive as individuals and as a collectivity.

A classic tribe may be tied to a specific territory and the exploitation of resources located there. Its formation may represent an evolution from the hunter-gatherer life of nomadic bands to a more settled agrarian village lifestyle. It may span various villages and hamlets, and its size may grow to several thousand people. Its identity as a tribe may harden as a result of conflicts with outsiders. And it may lack the formal institutional hierarchies that characterize chiefdoms and states, the two types of societies that come next in evolutionary theory. Yet even if these or other observations made by scholars are added to the definition of the tribe, kinship remains its essence.

As tribes grow, clans usually coalesce inside them. Clans are clusters of families and individuals who claim a particular lineage and, because of this, act conjointly in a corporate manner. Typically, a clan has its own legends, rituals, and ceremonies, its own lands, households, and other properties, a "Big Man" or an elder to represent (but not rule) it, and perhaps a particular function, such as progeny who serve as priests or warriors. Mutual defense and aid are keenly important in clan systems; indeed, an insult or threat to any one member is received as an insult or threat to all – as is also the case for a tribe as a whole vis-à-vis other tribes and outsiders.

While lineage and marriage ties can keep small tribes together, they alone do not suffice to keep large tribes and clans integrated. Integration on a larger scale requires the rise of a variant of the kinship principle: fraternal associations and corporate orders based more on a sense of brotherhood than on blood – what anthropologists call "fictive kinship." Such associations may combine individuals from various families and villages for a specific corporate purpose. Examples include secret brotherhoods and age-grade, warrior, healing, ceremonial, and religious associations. While some such brotherhoods may derive directly from lineage (e.g. a clan), others do not, yet all emulate kinlike relations. The larger and more complex a tribe becomes, the more important such brotherhoods become. (In modern times, these are often called clubs, gangs, and secret societies.)

Kinship considerations permeate everything – all thought and action – in a tribe and its constituent segments. One's identity is less about one's self than about one's lineage; lineage determines most of one's identity as an individual and submerges it in the tribal whole. This applies also to one of the most important activities in a tribe, namely, arranged marriage, which itself is about the linking of families, not individuals. From our distant remove, varied economic, political, and cultural activities may appear to occur in a tribe; but seen in their own light, tribes lack such differentiation; everything one does in a tribe is done as a kinsman of one kind or another. In tribal milieus, strategy and tactics revolve around what might be called *kinpolitik*, far more than realpolitik.

While kinship charts and calculations can become enormously complicated, it suffices to say here that individual identities and possibilities in tribal/clan societies are both fixed and fluid at the same time. Lineage positions are fixed, because of who one's parents were and when one was born. Moreover, as a rule, tribe trumps clan, family, and individuals, binding all into a nested social (but not political) hierarchy. Yet, kin and their associates operate on lateral as much as vertical ties; for example, a person can choose which relative (say, which distant cousin) to ally with on which issues and under what circumstances. This structure can make for highly flexible social possibilities that resemble not only circles within circles but also circles across circles. It offers extensive room for maneuver, which can be used for promoting rivalries as well as alliances.

As individuals, families, clans, and tribes as a whole assert their place and maneuver for position, maximizing honor – not power or profit – is normally their paramount motivation. This emphasis is often thought to flow from the fact that tribes arose in subsistence times, way too early for power or profit to matter. But there must be more to the explanation, for the pattern persists in modern varieties of tribes and clans. Wherever people (even powerful, rich people) turn tribal and clannish, honor and its concomitants – respect, pride, and dignity – come into serious play in social interactions. Thus, warlords and warriors fighting in Afghanistan, Iraq, and other tribal zones are renowned for the value they place on upholding codes of honor and avoiding shameful humiliation. Everyone wants to gain honor for themselves and their lineage, clan, or tribe; no one can afford to lose face, for that would reflect badly not only on them as individuals but also on all their kin. (If the word were in a dictionary, it might be said that tribes and clans are deeply "honoritarian.")

Let us turn next to organizational principles. Reflecting the primacy of kinship bonds, tribes are resolutely egalitarian, segmental, and acephalous – to use terms favored by anthropologists. These three principles are interlocking.

First, in being egalitarian, a tribe's members are deemed roughly equal to one another. The aim is not so much absolute equality as respect for individual autonomy, and especially the autonomy of individual households. In this spirit, members emphasize communal sharing, as in sharing food, giving gifts, and doing favors. These practices oblige recipients to reciprocate, for honorable reciprocity, not exchange, is the underlying ethic. Elitism is avoided, and domination efforts are not tolerated for long. Upstarts, such as alpha-type bullies and despotic self-aggrandizers, are eventually restrained, as are overly selfish free riders and oddball deviants. Indeed, classic tribes are so egalitarian that no fixed rank or status system exists in them. There are tendencies for elders to receive more respect than the young, men more than the women, and a "Big Man" more than others. Also, family heads may lord it over others inside their own households; and some lineages and clans may compete for status. But overall, the egalitarian ethos limits hierarchical and competitive tendencies. Whoever shows leadership has to be modest, generous, and self-effacing and treat others as peers. There is constant groupwide vigilance to keep anyone from gaining sway for long. If necessary, coalitions form to assure leveling. In tribal systems rent by feuds and rivalries, egalitarianism becomes more an ideal than a reality, but it is still the desired ethos. In short, tribes behave more like balance-of-honor than balance-of-power systems.

Second, the classic tribe is segmental, in that every part resembles every other – there is no specialization. Tribes have no distinct central nervous system, and all households and villages are essentially alike: resolutely self-sufficient and autonomous. Because tribes are so segmental and undifferentiated, their constituent parts – families, lineages, clans, and so on – tend to oscillate between fusion and fission. Fusion occurs, for example, when clan intermarriages foster unity across villages and other segments; when segments, even ones that were feuding, ally against an enemy; and when a tribe absorbs an outside band or tribe. Fission occurs when a tribe is so beset by shortages or feuds that a segment (e.g. a few related households, an entire clan) hives off and goes its own way, forming a new tribe that immediately replicates the design of the old. Whether in a state of fusion or fission, each segment guards its autonomy.

Third, the classic tribe is acephalous. The earliest form of social organization was not hierarchy; egalitarian tribes were the norm before hierarchical societies – first chiefdoms, then states – emerged. Classic tribes had no formal leaders, not even chiefs. Informal status differences that arose (e.g. deference to elders) were kept muted. Political hierarchies, dominant groups, class structures, and other status systems are absent at this stage. The title of chief, if there was one, meant little; a chief was a man of influence, an adviser, a facilitator, a broker – but he could not give orders that had to be obeyed. Thus, leadership, which might be needed when hunting for big game or in the conduct of a ceremony, was transient and low profile; it kept shifting and depended more on the situation than the person. One day's "Big Man" was not necessarily tomorrow's. Major decisions, such as whether to go to war or where to migrate, were made in tribal councils open to all, where anyone (at least all household heads) could speak. Indeed, consultative consensus seeking in tribal councils was the first form democracy took.

What matters for maintaining order and peace in such tribal milieus is not leadership, hierarchy, force, and law – it is too early a form for that – but the customs and codes of etiquette that flow from revering kinship bonds. Kinship systems place high value on principled, praiseworthy displays of respect, honor, trust, obligation, sharing, reciprocity, and acceptance of one's place. Rituals and ceremonies – and later, religion – reinforce these customs. In the event of wrongdoing, sanctions run the gamut from public blame, shame, shunning, ostracism, and a withdrawal of reciprocity to expulsion or execution if the group reaches consensus on it.

Principles of respect, dignity, pride, and honor are so important in a tribal society that humiliating insults may upset peace and order more than anything else. An insult to one individual is normally taken as an insult to all who belong to that lineage. There are only two ways to relieve the sense of injury from insult: compensation or revenge. A call for compensation or revenge may apply not just to the offending individual but to his or her entire lineage. Responsibility is collective. Justice is less about punishment for a crime than about gaining adequate compensation or revenge to restore honor. It is not unusual to find clans and tribes engaged in prolonged cycles of revenge and reconciliation – i.e. fission and fusion – deriving from insults that happened long ago.

These, in summary fashion and skipping many intricacies, are the basic dynamics of classic tribes. They took shape more than 5,000 years ago during Neolithic times. They characterize many bands, tribes, and some chiefdoms that social and cultural anthropologists have studied in recent eras, such as the Nuer (Africa), the Trobrianders (Melanesia), the !Kung (Africa), the Iroquois (North America), and the Yanomamo (Brazil), not to mention examples from European history. Some examples may look ancient, primitive, or backward. But the tribal form is not ancient history; it endures today – indeed, one manifestation or another makes media headlines almost every day. This is true not only for events in Africa, the Middle East, and South Asia but also in fully modern societies in North America and Europe, where the tribal paradigm is constantly reiterated in small but significant ways, such as in the often clannish organization and behavior of civic clubs, fellowships, fraternities, sports clubs (e.g. soccer hooligans), car clubs, and ethnic urban gangs, to note a few examples. All such organizations reflect the tribal paradigm, for they are normally more about ancient desires for identity, honor, and pride than about modern proclivities for power and profit.

War and religion in tribal settings

At its best, the tribal way of life imparts a vibrant sense of solidarity. It fills a people's life with pride, dignity, honor, and respect. It motivates families to protect, welcome, encourage, shelter, and care for each other (and guest outsiders) and to give gifts and hold ceremonies that affirm their connections to each other and to the ancestors, lands, and gods that define the tribe's identity. This kinship creates a stable realm of trust and loyalty in which one knows (and must uphold) one's rights, duties, and obligations. Many people around the world still prefer this ancient way of life over the ways of modern, impersonal, hierarchical and market systems. As noted above, even advanced societies that lack explicit tribes and clans still have tribelike sensibilities at their core that show up in nationalism, cultural festivities, civic interest groups, and sports and fan clubs.

But tribalism can make for a mean-spirited exclusivity and partiality too. Tribes and clans can be terribly sensitive about boundaries and barriers – about who is in the tribe and who is outside, about differences between "us" and "them." One's tribe (assuming it is not riven with feuds and rivalries) may seem a realm of virtue, where reciprocal altruism rules kin relations. But in tribal logic, virtuous behavior toward kin does not have to extend to others; outsiders can be treated differently, especially if they are "different."

Sometimes tribal exclusivity and partiality lead to war. When a tribe does go to war, it tries to do so as a whole, but it fights as segments. Internal feuds, rivalries, and other differences are set aside in order to unite against the outside enemy. Strategic agreement on the broad outlines of war may be reached in consultative councils. But each segment guards its own autonomy; not even in battle do they organize under a central command. If a war is based on alliances among groups within a tribe or between tribes, that may be another reason to guard autonomy; in tribal milieus, one day's ally may turn into another day's betrayer, and a group that takes shape one day may not be able to form anew later.

Classic tribal warfare emphasizes raids, ambushes, and skirmishes – attacks followed by withdrawals, without holding ground. Pitched battles are not the norm, for tribes lack the organizational and logistical capacities for campaigns and sieges. Sometimes the aims are limited, but tribal warfare often turns into total warfare, aimed at massacring an entire people, mercilessly. Killing women and children, taking women captive, torturing and mutilating downed males, scalping, and beheading are common practices. So is treachery, such as inviting people to a feast then slaughtering them on the spot. Tribal fighters may absorb women and children from an enemy tribe, but they do not hold prisoners. Enemies who are not massacred are put to flight, and their lands and homes seized. Bargaining in good faith to end a conflict becomes nigh impossible, for the attackers have denied the legitimacy of those whom they are attacking. In ancient times, this brutal way of war did not ease until the rise of chiefdoms and states, when leaders began preferring to subjugate rather than annihilate people. In today's world, examples are still easy to find – the Hutu massacres of Tutsis in Rwanda come readily to mind, as do episodes in the Balkans.

Tribes that go to war normally do so in the name of their god or gods. Indeed, many religions, from ancient totemism onward, have their deepest roots in tribal societies. The major monotheistic religions – Judaism, Christianity, and Islam – each arose in a tense tribal time in the Middle East. And each, in its oldest texts, contains passages that, true to traditional tribal ethics, advocate reciprocal altruism toward kin yet allow for terrible retribution against outside tribes deemed guilty of insult or injury. Today, centuries later, tribal and religious concepts remain fused in much of the world, notably Africa, the Middle East, and South Asia.

The more a religion commends the kinship of all peoples, the more it may lead to ecumenical caring across boundaries (as Islam often does). But the more a religion's adherents delineate sharply between "us" and "them," demonize outsiders and view their every kin (man, woman, child, combatant or noncombatant) as innately guilty, adhere to codes of revenge for touted wrongs, and seek territorial or spiritual conquests, all the while claiming to act on behalf of a deity, the more their religious orientation is utterly tribal, prone to violence of the darkest kind. This is as evident in the medieval Christian Crusades as in today's Islamic jihads, to mention only two examples.

All religious hatred – whether Christian, Jewish, Islamic, Buddhist, or Hindu – is sure to speak the language of tribe and clan, and that language is sure to be loaded with sensitivities about respect, honor, pride, and dignity, along with allocutions to the sacred, purifying nature of violence. This is a normal ethic of tribes and clans, whatever the religion. Indeed, as Amin Maalouf says about today's world in his *In the Name of Identity: Violence and the Need to Belong* (2001, pp. 28–9), "[I]f the men of all countries, of all conditions and faiths can so easily be transformed into butchers, if fanatics of all kinds manage so easily to pass themselves off as defenders of identity, it's because the 'tribal' concept of identity still prevalent all over the world facilitates such a distortion."

Savagery may worsen when tribal elements are led by a sectarian chieftain who is also a grandiose, ruthless warlord, like Osama bin Laden, Abu Musab al-Zarqawi, the Taliban's Mullah Muhammad Omar, or Chechnya's Shamil Basayev. If the outsiders they target (including Americans) react with a tribalism (or extreme nationalism) of their own, battles over whose religion should win become inseparable from those over whose tribe should win. While the modern idea of separating state and church is difficult enough, any notion of separating tribe and religion is inconceivable for many people, especially in wartime.

Al-Qaeda, Osama bin Laden, and global jihad

Osama bin Laden and al-Qaeda fit the tribal paradigm quite well. There is ample evidence that bin Laden thinks and operates in tribal/clan terms, as seen in his selection of wives, his aptitude for forming secretive brotherhoods, and his rhetoric about Islam, the Arab world, and jihad. The regions where al-Qaeda has been based are notoriously tribal: Afghanistan under the Taliban and now allegedly along the Afghan–Pakistan border. Also, al-Qaeda's main targets include Saudi Arabia, a tribal kingdom, and Iraq, where much of the population has reverted to tribal and clan ways since the collapse of the Iraqi state.

This is not the dominant way in which al-Qaeda and its affiliates are viewed. Analysts have preferred to keep looking for central decision-making nodes and specialized structures – even committees – for matters such as targeting, recruitment, financing, logistics, and communications, as though they might uncover a corporate pyramid. Or they have treated the creation of affiliates as though they were franchises that took the initiative to become affiliates or were concocted at al-Qaeda's behest. Or analysts have emphasized the sprawling network designs that al-Qaeda and its affiliates increasingly exhibit. Or they have applied social movement theory. All these analytical approaches make sense and should continue. But they end up making al-Qaeda look like a work of dauntless, modern, forward-looking genius, when it isn't. Its design looks backward more than it looks forward; it reiterates as much as it innovates – and that's because of its enduring tribalness.

The tribal paradigm – and a case that al-Qaeda is like a global tribe waging segmental warfare – shows up across five analytic dimensions: narrative content, social appeal, leadership style, organizational design, doctrine and strategy, and the use of information technology.

Narrative content

Many themes in bin Laden's and other jihadists' statements fit the tribal paradigm. The world is divided between good-hearted believers – the worldwide *umma* (kindred community) of Muslim brothers and sisters – and evil nonbelievers (infidels, apostates, heretics). Arab lands and peoples have suffered far too much injury, insult, and humiliation – their honor has been trampled, their families disrespected – by arrogant, self-aggrandizing intruders (particularly America and Israel). Muslims have a sacred duty to defend themselves, fight back, wreak vengeance, seek retribution, and oust the foreign invaders. The transgressors must be made to pay; no mercy should be shown – no matter if civilians die, even women and children. They deserve every punishment, every catastrophe, every tit-for-tat that can be heaped upon them. Defensive warfare is a necessary duty to restore honor and pride. This story line is made to sound Islamic, and it has Islamic aspects that are not necessarily tribal – for example, requiring that an enemy be warned. But overall, it is tribal to the core. Indeed, similar story lines have cropped up among virulently tribal Jewish, Christian, and other religious extremists as well, all across history.

Social appeal

Among Muslims, the jihad narrative is not alien, academic, or bizarre. It requires little indoctrination, for it arouses both the heart and mind. Recruits willingly come from among militants who fought in Afghanistan, Chechnya, or the Balkans; immigrants in Europe and refugees in Jordan and Palestine who are leading alienated, unsettled lives; youths leading comfortable but constricted lives in Saudi Arabia; and Sunnis whose lives have been shattered by the warring in Iraq. What drives them, according to many analyses, are shared sensibilities about loss, alienation, humiliation, powerlessness, and disaster. Such analyses may also note, more in passing than in depth, that joining al-Qaeda or an affiliate provides a familylike fellowship. However, this factor should not be given short shrift; participation may appeal largely because it binds members in such a fellowship – in mosques, training camps, militant cells, and so on. And it may do so not simply because many members share the social-psychological sensibilities noted above but because they come from cultures that are deeply, longingly tribal and clannish. For the lost and the adrift, joining al-Qaeda recreates the tribal milieu. This may even apply to the attraction of nomadic loners from faraway cultures who convert to Islam while seeking a more meaningful identity and sense of belonging for themselves (e.g. a John Walker Lindh).

Leadership style

Bin Laden's stylized demeanor is in the tradition of a modest, self-effacing, pious tribal sheik. He espouses, interprets, advises, facilitates, brokers, and blesses. His ideas are embedded in Islamic tradition – he does not concoct them to express his ego. He radiates a commanding presence, but he does not give orders or demand submission to his leadership (although he may well be chief of his own cell, i.e. household). He is generous with funds. His co-leader of al-Qaeda, Ayman al-Zawahiri, conveys a similar though edgier image. In contrast, their fellow warrior in Iraq, al-Zarqawi, acted like a ferocious alpha-male bully who would just as soon create fissions between Sunni and Shi'a tribes in Iraq. Yet he was so respectful of bin Laden, who takes a more ecumenical approach to pan-tribal fusion, that bin Laden declared him to be his emissary in Iraq. Information is lacking on how these and other chieftains make decisions affecting al-Qaeda, but the process appears to involve mutual communication, consultation, and accommodation to reach a consensus that does not smack of hierarchy or imposition – much as might occur in a classic tribal council.

Organizational design

Al-Qaeda and its affiliates are organized as a (multihub? core/periphery?) network of dispersed nodes, cells, and units, all campaigning in a similar direction without a precise central command. This structure looks like an information-age network, but it is equally a tribal-age network. It is bound together by kinship ties of blood and especially brotherhood. What look like nodes and cells from a modern perspective correspond to segments from a tribal perspective. Some segments come from true tribes and families; others are patched together in terms of "fictive kinship" by jihadist clerics, recruiters, and trainers. Yet all who join get to feel like they belong to segments of an extended family/tribe that reaches around the world. Al-Qaeda had a segmentary quality even before 9/11; for example, some training camps in Taliban Afghanistan were divided along ethnic lines (e.g. here for Algerians, there for Chechens), and the cells that struck on 9/11 consisted of a Saudi segment. Furthermore, this jihadist network is vaguely acephalous (or polycephalous), as a tribe should be. It is held together not by command-and-control structures – tribes are not command-and-control systems – but by a gripping sense of shared belonging, principles of fusion against an outside enemy, and a jihadist narrative so compelling that it amounts to both an ideology and a doctrine.

Doctrine and strategy

Al-Qaeda and its affiliates fight in the field much like tribes and clans: as decentralized, dispersed, semiautonomous segments that engage in hit-and-run (and hit-and-die) tactics. These segments vary in size and makeup. Some are small and fit our notion of terrorist cells. Others (e.g. in Afghanistan and Iraq) are larger, more like platoons with commanders (thus it might be more accurate to refer to Ba'thist segments than Ba'thist cells). Some may resemble close-knit, exclusive brotherhoods; others may keep shifting in membership. Meanwhile, they fight like modern terrorists and insurgents but do so in the tradition of tribal warriors, relying on stealth, surprise, treachery, and savagery while avoiding pitched battles. And they are comfortable with temporary marriages of convenience, such as in Iraq, where Ba'thist and Islamist units cooperate on tactical missions but keep separate organizations and strategies. The absence of a central hierarchy is not a sign of disorganization or weakness – it is the tribal way. Thus, while al-Qaeda's underlying doctrine and strategy have been acquiring the sophistication of modern notions of asymmetrical warfare (e.g. for netwar and swarming), its tribalness endures within that modern frame.

Use of information technology

Al-Qaeda and its affiliates have an extensive, growing presence on the Internet. Their statements, speeches, and videos are posted on myriad Websites around the world that advocate, sympathize with, and report on jihad. As many analysts have noted, the new information media are enabling terrorists and insurgents to augment their own communication and coordination and to reach outside audiences. The online media also suit the oral traditions that tribal peoples prefer. What merits pointing out here is that the jihadis are using the Internet and the Web to inspire the creation of a virtual global tribe of Islamic radicals – an online umma with kinship segments around the world. The Internet can help members keep in touch with a segment or reattach to a new segment in another part of the world as they move around. Thus the information revolution, not to mention broader aspects of globalization, can facilitate a resurgence of intractable tribalism around the world. Al-Qaeda and its ilk are a leading example.

In other words, al-Qaeda is like a global tribe, waging a modernized kind of segmental (or segmented) warfare. In Afghanistan and Iraq, we are fighting against virulent tribalism as much as Islamic fundamentalism. Salafi and Wahhabi teachings urging jihad against infidels, fatwas issued by Islamic sheiks to justify murdering even noncombatants, and stony ultimatums from Sunni insurgents who behead captives are all more manifestations of extreme tribalism than of Islam. In Islam, jihad is a religious duty. But the interpretation of jihad that al-Qaeda practices is rooted less in religion than in the appeal of virulent tribalism in some highly disturbed contexts.

Overlap with the network paradigm

American analysts and strategists should be treating al-Qaeda more as a tribal than a religious phenomenon. They should be viewing al-Qaeda from the classic tribal as well as the modern network perspective. It is often pointed out (including by me) that al-Qaeda represents a postmodern, information-age phenomenon. But it is time to balance this view with a recognition that al-Qaeda also represents a resurgence of tribalism that is both reacting to and taking advantage of the information revolution and other aspects of globalization.

The tribal view overlaps with the invaluable network view of al-Qaeda, particularly the one that Arquilla and I have called "netwar" (Arquilla and Ronfeldt 2001), in which the protagonists use network forms of organization and related doctrines, strategies, and technologies attuned to the information age. Netwar protagonists – like al-Qaeda and its affiliates – tend to consist of dispersed groups and individuals who communicate, coordinate, and act conjointly in an internetted manner, often without a central command. Their optimal mode of attack is stealthy swarming. In many respects, the netwar design, like al-Qaeda's, resembles the "segmented, polycentric, ideologically integrated network" (or SPIN) that Luther Gerlach (1987) identified in his 1960s studies of social movements.

But tribes and networks are not the same phenomenon. For one thing, tribes are ruled by kin relations, whereas information-age networks operate mainly according to modern criteria. Consider the issue of information sharing, for example. In tribal systems, the sharing of information may proceed after a recipient's lineage has been checked and found suitable. In networks, the decision criteria are not about lineage but the professional nature of the role or person who may receive the information. Also, in tribal and clan systems in which members are maneuvering for influence, fluid alliances often arise that look odd and contradictory to outsiders from an ideological or other modern perspective but are sensible from a tribal or clan perspective. For example, it may behoove a tribal or clannish elite circle (as in the old Iranian *dowreh* or Mexican *camarilla* systems) to stealthily include elites from right and left, military and religious, business and criminal sectors, so that the circle is plugged into all circuits vying for position in a society. In contrast, modern networks – for example, in the area of civil-society activism – generally aim for ideological and professional coherence. Finally, if tribes and networks were similar concepts (as some social network analysts might argue), then modern corporations might as well be advised to adapt to the information revolution by becoming more tribal instead of networked – but that is patently not sensible, except where particular issues are concerned, such as employee morale or product branding.

In short, al-Qaeda and its affiliates have formed a hybrid of the tribal and network designs – a tribalized network or networked tribe, so to speak, that includes bits of hierarchy and marketlike dynamics as well. The tribal paradigm has a striking advantage over the network, hierarchy, and other organizational paradigms. The latter models point to organizational design first, and then to matters of leadership, doctrine, and strategy. But they have nothing clearly embedded in them about religion. As voiced in terrorism discussions, they are secular paradigms; religion is grafted on, as a separate matter. In contrast, the tribal paradigm is inherently fraught with dynamics that turn into religious matters, such as altruism toward kin, delineations between "us" and "them," and codes of revenge. And that is another valuable reason to include the tribal paradigm in analyses of al-Qaeda and other terrorist movements.

Preliminary implications for policy and strategy

Americans comfort themselves by thinking that no other nation will be able to match our power for decades to come. But from ancient times to the present, great powers that expand globally often run into subnational tribes or clans that resist fiercely, even unfathomably. Sometimes this has dire, wasting consequences (as it did for the Roman Empire), although a great power can extemporize by playing segments against each other (as Britain did during the Pax Britannica). Also, ever since ancient times, the more tribal or clannish a society, the more resistant it is to change and the more pressures for modernizing reforms must come largely from outside or above (e.g. Meiji Japan). Americans still have much to learn about dealing with tribalized and clannish societies and devising programs that work in them (remember Somalia).

The United States is not at war with Islam. Our fight is with terrorists and insurgents who are operating in the manner of networked tribes and clans. US military forces are learning this the hard way – on the ground. But policymakers and strategists in Washington still lag in catching on. For example, the *Report of the Defense Science Board Task Force on Strategic Communication* (Defense Science Board 2004, p. 2) recognizes quite sensibly that "the United States is engaged in a generational and global struggle about ideas, not a war between the West and Islam." It notes the role of tribalism, but only barely. A RAND report titled *The Muslim World After 9/11* (Rabasa *et al.* 2004, p. xx) goes further in saying that "extremist tendencies seem to find fertile ground in areas with segmentary lineal tribal societies," but it mainly laments the fact that "the literature on the relationship between tribalism and radicalism is not yet well developed."

US counterinsurgency and counterterrorism methods – for interrogations, intelligence assessments, information operations, strategic communications, and public diplomacy, and indeed, for the whole "war of ideas" – would benefit from our upgrading our understanding of tribal and clan dynamics. Although identifying exactly what reconsiderations should take hold is beyond the scope of this paper, generally speaking, we must learn better ways to separate our strategies toward Islam from our strategies toward tribalized extremists who ultimately cannot endure such a separation. Whose story wins may well depend largely on just that.

The tribal paradigm may be useful for rethinking not only how to counter al-Qaeda but also what may lie ahead if al-Qaeda or an affiliate ever succeeds in seizing power and installing an Islamic caliphate somewhere. Then neither the tribal nor the network paradigm would continue to be so central. Hierarchy would come to the fore as a caliphate is imposed. Over the ages, people have come up with four major forms of organization for constructing their societies: tribes, hierarchical institutions, markets, and networks. How people use and combine these forms, both their bright and dark sides, pretty much determines what kind of society they have. Were an al-Qaeda-inspired caliphate to take root, we can be pretty sure that it would combine hyperhierarchy and hypertribalism while leaving marginal, subordinate spaces for economic markets and little if any space for autonomous civil society networks. When this has occurred in the past, the result has typically been fascism.

Postscript: redirecting the "war of ideas" (May 2006)

Events since an earlier version of the preceding text was published in 2005 have further substantiated the thesis that al-Qaeda-type terrorism involves a strong dose of tribalism, in all the ways noted above. But analysts and strategists continue to have difficulty using a tribal paradigm. In this postscript, I stress the theme anew and elaborate a little more on the "so what" question for information strategy.

Neglect of the tribal paradigm by terrorism analysts and grand strategists

The significance of tribalism has not been ignored by analysts and strategists. It just has not gained traction. First of all, the tribalism theme has never been out of reach for terrorism analysts. Most are aware, for example, that terrorism often involves a "true believer" mentality. The originator of "true believer" analysis, Eric Hoffer, warned about its tribal nature decades ago:

> To ripen a person for self-sacrifice he must be stripped of his individual identity and distinctness.... The most drastic way to achieve this end is by the complete assimilation of the individual into a collective body.... When asked who he is, his automatic response is that he is a German, a Russian, a Japanese, a Christian, a Moslem, a member of a certain tribe or family. He has no purpose, worth and destiny apart from his collective body; and as long as that body lives he cannot really die....
>
> This is undoubtedly a primitive state of being, and its most perfect examples are found among primitive tribes. Mass movements strive to approximate this primitive perfection, and we are not imagining things when the anti-individualist bias of contemporary mass movements strikes us as a throwback to the primitive.
>
> (Hoffer 1989 [1951], pp. 62–3)

These are solid points; and indeed, the worst terrorist attacks in the West – those on 11 September 2001 in New York and Washington, on 11 March 2004 in

Madrid, and on 7 July 2005 in London – were mounted in part by Islamic "true believers." But while this concept continues to receive occasional notice, Hoffer's diagnosis of its tribal nature is rarely noted.

Most analysts prefer mainstream psychological, social, and cultural frameworks for working on terrorism. They also prefer terms such as "sectarian" or "factional." These frameworks and terms are not inaccurate, but they neglect explicitly identifying the tribal and clan dynamics that, according to my analysis above, lie at the heart of organized terrorism. (It is not only al-Qaeda but also other terrorist movements – notably, all the segments that have fought and feuded in Ireland – that may be reinterpreted from a tribalism perspective. So may violent, clannish urban youth gangs – such as the Bloods and Crips in Los Angeles and the transnational Mara Salvatrucha.)

Second, the tribalism theme has never been far out of reach for grand strategists either – but again it has not gained traction. Broader tendencies in American strategic thinking have not helped. To begin with, a leading new idea after the collapse of the Soviet Union and the end of the Cold War held that history would progress next to fulfill liberal democratic ends – the "end of history" argument (Fukuyama 1989, 1992). This was a valuable idea, but it was too premature and optimistic to hold center stage for long. It looked too far forward; most of the world was not suited to it yet.

To other thinkers at the same time, a contrary idea – regression to tribalism – made greater sense, particularly as murderous ethnonationalist hatred unfolded in freed-up parts of Eastern Europe. For example, one observer warned:

> Unless we take early action, I fear that Yugoslavia will be only a beginning. A new variant of the domino theory will be upon us. But the dominoes will fall this time not toward communism, not even toward nationalism, but rather toward tribalism.... Without action to stop wholesale disintegration the basic political unit will become the tribe, its normal state one of war with all other tribes, each fighting for a position of dominance it can only hold temporarily, each feeling insecure and threatened by its neighbors, and each seeking alliances to pursue its narrow ends.
>
> (Attali 1992, p. 40)

This view recognized that where societies crumble under great stress, people are likely to revert to tribal and clan behaviors that repudiate liberal ideals. However, for whatever reason – perhaps because "tribalism" sounds too archaic and anthropological, or because it was voiced more by European than American strategists, or because it was never spelled out authoritatively in detail – it did not take hold. Instead, grand strategic thinking and policy dialogue came to revolve around a more high-minded but less accurate concept that soon emerged: "the clash of civilizations" (Huntington 1993/1996).

Now, over a decade later, this mostly pessimistic concept remains central to the "war of ideas" around the world. It is voiced at times not only by Western political leaders, such as British prime minister Tony Blair, but also by Islamists in the Middle East and South Asia, notably during protests against the Danish cartoons depicting the Prophet Muhammad in early 2006. The clash of civilizations is not the only idea in play among policymakers and strategists – promotion of democracy still matters – but it has acquired a kind of grip on grand narratives about where the world is headed, as expressed in story lines such as "the West versus Islam." Tribalism is treated as a secondary or tertiary phenomenon.

If the tribal paradigm were taken more seriously by terrorism analysts and global strategists, it could affect all the areas noted briefly earlier: interrogations, intelligence, information operations, strategic communications, public diplomacy, indeed the whole "war of ideas" – all that falls under the broad rubric of information strategy, and what Arquilla and I (1999) call "noopolitik." Of these matters, I limit my remarks in this postscript to the "war of ideas" and some of its implications for information strategy.

The crucial clash: a turmoil of tribalisms

The "war of ideas" against al-Qaeda and its affiliates is said to involve a "test of wills" and a "battle of beliefs" for the purpose of "winning hearts and minds." The language that keeps rising to the top in the West portrays this clash as a war about freedom and tyranny, good and evil, innocence and guilt. Thus, it is about who is and is not civilized and whether Islam is being hijacked by savage extremists. But is this an optimal basis for information strategy?

It is not. The catchy phrases about civilization are working poorly and proving to have inherent limitations. If it is said that the fight against al-Qaeda (not to mention the insurgents in Iraq) fits the clash-of-civilizations paradigm, then by implication the terrorists (and insurgents) somehow represent a civilization. Moreover, if the fight is said to be between the forces of civilization (mostly meaning the West) and its barbaric opponents (meaning Islamic terrorists), then a rhetorical door is left open for counterclaims that the West is not as civilized as it supposes and that Islam (not to mention the Arab world) has contributed greatly to the growth of civilization for over a thousand years. Besides, all great civilizations, as they arise, display potent new combinations of military might and religious zeal – al-Qaeda and its ilk can claim that this is what they are doing and set heads nodding among sympathizers who yearn for the revival of a great Islamic caliphate.

Thus, it is inadvisable for American strategists to continue playing up, or playing to, the clash-of-civilizations perspective. It does not provide an accurate, appealing, or agile basis for information strategy. In some respects, it plays into the cognitive hands – the "hearts and minds" – of the terrorists and their sympathizers.

What troubles the world today is far more a turmoil of tribalisms than a clash of civilizations. The major clashes around the world are not between civilizations per se but between antagonistic segments that are fighting across fringe border zones, such as Christian Serbs and Muslim Kosovars, or feuding within the same civilizational zone, such as Sunnis and Shi'ites in Iraq. And most such antagonists, no matter how high-mindedly they proclaim their ideals, are operating in terribly tribal and clannish ways. Some, including al-Qaeda terrorists, are extreme tribalists who may dream of making the West start over at a razed tribal level.

Indeed, essentially tribal or clannish societies, chiefdoms, and clan-states, some dressed in the trappings of nation-states and capitalist economies, remain a ruling reality in much of the world. The densest arc of tribalism runs across North Africa, the Middle East, and South Asia and up into the "-stans" of Central Asia. There, with exceptions (e.g. Egypt, Israel, Iran, and India), what pass for nation-states mostly amount to hardy hybrids of tribes, clans, chiefdoms, kingdoms, and feudal gangster regimes. All across this arc, the observation applies that "much of the warfare today is more like past tribe- and chiefdom-level warfare than the high-tech wars of the modern era" (LeBlanc 2003, p. xv). Yet, this is where US power is being most fiercely invested and contested these days.

This turmoil of tribalisms is sure to persist a long time, fueling terrorism, ethnonationalism, religious strife, sectarian feuds, clannish gang violence, and crime. The cartoon-related riots by Muslims in 2006 fit this pattern; they represented an endeavor to mobilize a global tribe, not a civilization. And even though al-Qaeda and its affiliates constitute an information-age network, they too continue to operate like a global tribe – decentralized, segmental, lacking in central hierarchy, egalitarian toward kith and kin, ruthless toward others.

Besides being more accurate, the tribalism paradigm also better illuminates the crucial problem: the tribalization of religion. As discussed earlier, the more these extremists foster prejudicial divisions between "us" and "them," vainly claim sacredness solely for their own ends, demonize others, revere archaic codes of revenge, crave territorial and spiritual conquests, and suppress moderates who disagree – all the while claiming to act on behalf of a deity – the more their religious orientation is utterly tribal and prone to wreaking violence of the darkest kind. They can only pretend to represent a civilization.

Whose story wins: extremists or moderates?

In short, Islam, a civilizing force, has fallen under the spell of Islamists who are a tribalizing force. In the war of ideas, as well as in the battles on the ground, whose story wins may depend largely on addressing this brand of tribalization.

Shifting to a turmoil-of-tribalisms perspective would have to be carefully thought out. The point is not to condemn all tribal ways. Many people around the world appreciate (indeed prefer) this communal way of life and will defend it from insult. Moreover, even the most modern societies retain tribal tendencies at their core – as expressed, for example, in nationalism, cultural pride, and all sorts of civic groups and fan clubs that express social identities. That must be upheld; it is not always uncivilized to be tribal. Instead, the point is to strike at the awful effects that extreme tribalization can have – to oppose not the terrorist's or insurgent's religion but rather the reduction of that religion to raw tribalist tenets.

This approach could help rally moderates to resist clannish, sectarian extremists. Western leaders have put Muslim leaders everywhere under pressure to denounce terrorism as barbaric and uncivilized. But this approach to the "war of ideas," along with counterpressures from sectarian Islamists, has put moderate Muslims on the defensive, often inhibiting them from speaking out. An approach that focuses instead on questioning extreme tribalism, particularly the tribalization of religion, may be more effective in freeing up dialogue and inviting a search for common ecumenical ground.

This is a domain – and a task – for information strategy. It involves knowing the enemy, shaping public consciousness, and crafting persuasive messages for friend and foe alike. It involves getting the content of those messages right and finding the best conduits for them. It is about winning the battle of the story. And it involves doing all this in such a way that soft power works as well as hard power, and information-age *noopolitik* outperforms traditional realpolitik.

Sources

The sections on classic tribal dynamics and tribal warfare draw heavily on the sources listed below. Many points are condensed and paraphrased from them, and there is hardly an idea or observation in those sections that does not come from those sources. In addition to the sources listed below, some points derive from articles in the *New York Times* and the *Los Angeles Times* and from a C-SPAN2 broadcast of the conference "Al Qaeda 2.0: Transnational Terrorism After 9/11," convened by the New America Foundation and the New York University Center on Law and Security in Washington, DC, 2 December 2004, which included presentations by many top experts in the field of terrorism.

Sources on classic tribes

Boehm, Christopher, *Hierarchy in the Forest: The Evolution of Egalitarian Behavior*, Cambridge, MA: Harvard University Press, 1999.

Burguière, André, Christiane Klapisch-Zuber, Martine Segalen, and Françoise Zonabend, eds, *A History of the Family*, 2 vols, translated from 1986 French edition, Cambridge, MA: Belknap Press of Harvard University Press, 1996.

Carneiro, Robert L., *Evolutionism in Cultural Anthropology: A Critical History*, Boulder, CO: Westview Press, 2003.

Earle, Timothy, *How Chiefs Come to Power: The Political Economy in Prehistory*, Stanford: Stanford University Press, 1997.

Evans-Pritchard, E. E., *The Nuer: A Description of the Modes of Livelihood and Political Institutions of a Nilotic People*, Oxford, UK: Oxford University Press, 1940.

Fargues, Philippe, "The Arab World: The Family as Fortress," in Burguière *et al.*, eds, *A History of the Family*, vol. 2, Cambridge, MA: Belknap Press of Harvard University Press, 1996, pp. 339–74.

Fox, Robin, *Kinship and Marriage: An Anthropological Perspective*, Cambridge, UK: Cambridge University Press, 1967 (reissued with new preface, 1983).

Fried, Morton H., *The Evolution of Political Society: An Essay in Political Anthropology*, New York: Random House, 1967.

Fried, Morton H., *The Notion of Tribe*, Menlo Park, CA: Cummings, 1975.

Harris, Marvin, *Cannibals and Kings: The Origin of Cultures*, New York: Random House, 1977 (reissued: Vintage Books, 1991).

Johnson, Allen W. and Timothy Earle, *The Evolution of Human Societies: From Foraging Group to Agrarian State*, Stanford: Stanford University Press, 1987.

Malinowski, Bronislaw, *Argonauts of the Western Pacific: An Account of Native Enterprise and Adventure in the Archipelagoes of Melanesian New Guinea*, Long Grove, IL: Waveland Press, 1984 (originally published in 1922).

Mauss, Marcel, *The Gift: The Form and Reason for Exchange in Archaic Societies*, transl. W. D. Halls, New York: W. W. Norton, 1990 (originally published in France in 1950).

Maybury-Lewis, David, *Millennium: Tribal Wisdom and the Modern World*, New York: Viking Penguin, 1992.

Sahlins, Marshall D., *Tribesmen*, Englewood Cliffs, NJ: Prentice-Hall, 1968.

Sanderson, Stephen K., *Social Transformations: A General Theory of Historical Development*, expanded edn, New York: Rowman & Littlefield, 1999.

Sanderson, Stephen K., *The Evolution of Human Sociality: A Darwinian Conflict Perspective*, New York: Rowman & Littlefield, 2001.

Schneider, David M., *American Kinship: A Cultural Account*, 2nd edn, Chicago: University of Chicago Press, 1980.

Service, Elman R., *Primitive Social Organization: An Evolutionary Perspective*, 2nd edn, New York: Random House, 1971.

Service, Elman R., *Origins of the State and Civilization: The Process of Cultural Evolution*, New York: W. W. Norton, 1975.

Shermer, Michael, *The Science of Good and Evil: Why People Cheat, Gossip, Care, Share, and Follow the Golden Rule*, New York: Henry Holt, 2004.

Tiger, Lionel, and Robin Fox, *The Imperial Animal*, New Brunswick, NJ: Transaction Publishers, 1998.

Wolfe, Alvin W., "Connecting the Dots Without Forgetting the Circles, *Connections*, vol. 26, no. 2 (2005): 107–19 (available at www.insna.org/Connections-Web/Volume26–2/10.Wolfe.pdf).

Sources on tribal and clan warfare

Ferguson, R. Brian, "Archaeology, Cultural Anthropology, and the Origins and Intensifications of War," in E. Arkush and M. Allen, eds, *The Archaeology of Warfare: Prehistories of Raiding and Conquest*, Gainesville: University Press of Florida, 2006.

Ferguson, R. Brian, "Tribal, 'Ethnic,' and Global Wars," in Mari Fitzduff and Chris Stout, eds, *The Psychology of Resolving Global Conflicts: From War to Peace*, vol. 1, *Nature vs. Nurture*, Westport, CT: Praeger Security International, 2006, pp. 41–69.

Furnish, Timothy, "Beheading in the Name of Islam," *Middle East Quarterly*, vol. 12, no. 2 (Spring 2005): 51–7.

Galeotti, Mark, "'Brotherhoods' and 'Associates': Chechen Networks of Crime and Resistance," *Low Intensity Conflict and Law Enforcement*, vol. 11, no. 2/3 (Winter 2002): 340–52.

GlobalSecurity.Org: "Iraq: Societal Framework," as modified 22 June 2005 (online only at www.globalsecurity.org/military/world/iraq/society.htm).

Keeley, Lawrence H., *War Before Civilization: The Myth of the Peaceful Savage*, New York: Oxford University Press, 1996.

LeBlanc, Steven A., with Katherine E. Register, *Constant Battles: The Myth of the Peaceful, Noble Savage*, New York: St Martin's Press, 2003.

McCallister, William S., "The Iraq Insurgency: Anatomy of a Tribal Rebellion," *First Monday*, vol. 10, no. 3 (March 2005) (online only at www.firstmonday.org/issues/issue10_3/mac/index.html).

Otterbein, Keith F., *How War Began*, College Station: Texas A&M University Press, 2004.

Simons, Anna, *Networks of Dissolution: Somalia Undone*, Boulder, CO: Westview Press, 1995.

Taylor, Richard L., *Tribal Alliances: Ways, Means, and Ends to Successful Strategy*, Carlisle, PA: Strategic Studies Institute, US Army War College, August 2005 (available at www.strategicstudiesinstitute.army.mil/pubs/display.cfm?PubID=619).

Other sources

Almond, Gabriel A., R. Scott Appleby, and Emmanuel Sivan, *Strong Religion: The Rise of Fundamentalisms Around the World*, Chicago: University of Chicago Press, 2003.

Anonymous (Michael Scheuer), *Through Our Enemies' Eyes: Osama bin Laden, Radical Islam, and the Future of America*, Washington, DC: Brassey's, 2002.

Arquilla, John and David Ronfeldt, *The Advent of Netwar*, Santa Monica, CA: RAND, 1996.

Arquilla, John and David Ronfeldt, *The Emergence of Noopolitik: Toward an American Information Strategy*, Santa Monica, CA: RAND, 1999.

Arquilla, John and David Ronfeldt, *Swarming and the Future of Conflict*, Santa Monica, CA: RAND, 2000.

Arquilla, John and David Ronfeldt, *Networks and Netwars: The Future of Terror, Crime, and Militancy*, Santa Monica, CA: RAND, 2001.

Arquilla, John and David Ronfeldt, "Netwar Revisited: The Fight for the Future Continues," *Low Intensity Conflict and Law Enforcement*, vol. 11, no. 2/3 (Winter 2002): 178–89.

Attali, Jacques, "Europe's Descent into Tribalism," *New Perspectives Quarterly* (Fall 1992): 38–40.

Barber, Benjamin R., *Jihad vs. McWorld: How Globalism and Tribalism Are Reshaping the World*, New York: Ballantine Books, 1996.

Bill, James A., "The Plasticity of Informal Politics: The Case of Iran," *Middle East Journal* (Spring 1973): 131–51.

Borgatti, Steven P. and Martin G. Everett, "Models of Core/Periphery Structures," *Social Networks*, no. 21 (1999): 375–95 (available at www.analytictech.com/borgatti/CP_Structure.doc).

Brimley, Shawn and Aidan Kirby, "Al Qaeda's virtual sanctuary," *Toronto Star*, 23 August 2005, p. A18.

Carley, Kathleen M., Jeffrey Reminga, and Natasha Kanmeva, "Destabilizing Terrorist Networks," North American Association for Computational Social and Organization Sciences conference proceedings, Pittsburgh, PA, 2003 (available at www.casos.cs.cmu.edu/publications/resources_others/a2c2_carley_2003_destabilizing.pdf).

Davis, Paul K. and John Arquilla, *Thinking About Opponent Behavior in Crisis and Conflict: A Generic Model for Analysis and Group Discussion*, Santa Monica, CA: RAND, 1991.

Defense Science Board, *Report of the Defense Science Board Task Force on Strategic Communication*, Washington, DC: Office of the Under Secretary of Defense for Acquisition, Technology, and Logistics, September 2004 (available online via www.acq.osd.mil/dsb/reports.htm).

Fukuyama, Francis, "The End of History?" *National Interest*, vol. 16 (Summer 1989): 3–18.

Fukuyama, Francis, *The End of History and the Last Man*, New York: Free Press, 1992.

Gerlach, Luther P., "Protest Movements and the Construction of Risk," in B. B. Johnson and V. T. Covello, eds, *The Social and Cultural Construction of Risk*, Boston: D. Reidel, 1987, pp. 103–45.

Hoffer, Eric, *The True Believer: Thoughts on the Nature of Mass Movements*, New York: Harper Perennial, 1989 (originally published in 1951).

Hoffman, Bruce, "Al Qaeda and the War on Terrorism: An Update," *Current History* (December 2004): 423–7.

Hoffman, Bruce, *The Use of the Internet by Islamic Extremists*, Santa Monica, CA: Rand, May 2006.

Huntington, Samuel P., *The Clash of Civilizations and the Remaking of World Order*, New York: Simon & Schuster, 1996.

Juergensmeyer, Mark, *Terror in the Mind of God: The Global Rise of Religious Violence*, 3rd edn, revised and updated, Berkeley: University of California Press, 2003.

Krebs, Vladis E., "Uncloaking Terrorist Networks," *First Monday*, vol. 7, no. 4 (April 2002) (online only at www.firstmonday.org/issues/issue7_4/krebs/index.html).

Krebs, Valdis and June Holley, "Building Sustainable Communities Through Network Building," 2002 (available at www.orgnet.com/BuildingNetworks.pdf).

Maalouf, Amin, *In the Name of Identity: Violence and the Need to Belong*, transl. Barbara Bray, New York: Arcade Publishing, 2001.

Nye, Joseph S., *Soft Power: The Means to Success in World Politics*, New York: Public Affairs, 2004.

Paxton, Robert O., *The Anatomy of Fascism*, New York: Alfred A. Knopf, 2004.

Post, Jerrold M., "When Hatred Is Bred in the Bone: Psycho-cultural Foundations of Contemporary Terrorism," *Political Psychology*, vol. 26, no. 4 (2005): 615–36.

Rabasa, Angel M., Cheryl Benard, Peter Chalk, C. Christine Fair, Theodore Karasik, Rollie Lal, Ian Lesser, and David Thaler, *The Muslim World After 9/11*, Santa Monica, CA: RAND, 2004.

Robb, John, "Global Guerrillas: Networked tribes, infrastructure disruption, and the emerging bazaar of violence: An open notebook on the first epochal war of the 21st Century," blog since November 2004, located at globalguerrillas.typepad.com/ globalguerrillas/.

Rodríguez, José A., "The March 11th Terrorist Network: In Its Weakness Lies Its Strength," Working Paper WP EPP-LEA: 03, Power and Privilege Studies Group, Department of Sociology, University of Barcelona, Spain, December 2005 (available at www.ub.es/epp/redes/redes.htm).

Ronfeldt, David, *Tribes, Institutions, Markets, Networks: A Framework About Societal Evolution*, Santa Monica, CA: RAND, 1996 (available at www.rand.org/publications/P/P7967).

Ronfeldt, David, "A Long Look Ahead: NGOs, Networks, and Future Social Evolution," in Robert Olson and David Rejeski, eds, *Environmentalism and the Technologies of Tomorrow: Shaping the Next Industrial Revolution*, Washington, DC: Island Press, 2005, pp. 89–98 (available at www.rand.org/publications/RP/RP1169/).

Ronfeldt, David, "Al Qaeda and Its Affiliates: A Global Tribe Waging Segmental Warfare?" *First Monday*, vol. 10, no. 3 (March 2005) (online only at www.firstmonday.org/issues/issue10_3/ronfeldt/index.html).

Ronfeldt, David, "Today's Wars Are Less About Ideas Than Extreme Tribalism," *Christian Science Monitor*, March 27, 2006.

Ronfeldt, David, and John Arquilla, "Networks, Netwars, and the Fight for the Future," *First Monday*, vol. 6, no. 10 (October 2001) (online only at www.firstmonday.org/issue6_10/index.html).

Stern, Jessica, *Terror in the Name of God: Why Religious Militants Kill*, New York: HarperCollins, 2003.

Tsvetovat, Maksim, and Kathleen M. Carley, "Structural Knowledge and Success of Anti-Terrorist Activity: The Downside of Structural Equivalence," *Journal of Social Structure*, vol. 6, no. 2 (2005) (online only at www.cmu.edu/joss/content/articles/volume6/TsvetovatCarley/index.html).

Weimann, Gabriel, *Terror on the Internet: The New Arena, the New Challenges*, Washington, DC: United States Institute of Peace, 2006.

3   Winning hearts and minds

A social influence analysis

Anthony R. Pratkanis

Before the French Revolution, most wars in Europe, according to Clausewitz (1832/1984), were fought by professional armies of limited size and interests. Rulers were generally unable to involve their subjects directly, and therefore once their forces had been defeated, it was difficult to mobilize a new army. Thus, the devastating effects of war tended to be minimized. However, beginning with Napoleon, war "became the business of the people" (Clausewitz 1832/1984, p. 592), and the people's passions and energy were stoked and mobilized through rhetoric and propaganda (Holtman 1950). The course of war was no longer determined by a rational calculus of interests of elite rulers but by the prejudices and emotions of everyday people, as evidenced in the US Civil War, two world wars, the Cold War, and numerous ethnic conflicts and acts of terrorism. It has become an accepted principle that to win a modern war – or any other conflict, for that matter, since "war is merely the continuation of politics by other means" – one must win what has been called "hearts and minds" (Arquilla and Ronfeldt 1999).

But how does one go about winning hearts and minds, and what exactly does it mean to win them? One approach, known as "soft power," seeks to promote the interests of the United States though attraction as opposed to coercion; soft power is the use of the attractiveness of a country's culture, political ideals, and policies to get others to admire those ideals and then follow one's lead (Nye 2004; Nye and Owens 1996). The sources of US soft power are its popular culture, which is enjoyed throughout the world, and the values underlying its domestic and foreign policies. This approach to winning hearts and minds uses such devices as public diplomacy, pamphlets, Voice of America, and other means to explain US policy; cultural exchange programs, especially with emerging leaders of other countries, to create an appreciation and understanding of American culture; and the sponsorship of American studies programs in overseas schools and universities (Ross 2002).

While there is nothing wrong with these devices per se, even a cursory look at the information campaigns conducted during the conflicts and wars of the nineteenth and twentieth centuries reveals that the objective of such campaigns were not just to create a positive cultural attraction, but to change the mind and behavior of enemies, neutrals, and those who might enter alliances. These campaigns had such objectives as winning support for ideals (e.g. free vs. slave; democracy vs. fascism and communism); creating and disrupting strategic alliances; countering the propaganda and deception of dictators, tyrants, and terrorists; stopping ethnic genocide and cleansing (and sadly, in many cases, promoting it); encouraging belligerents to cease fire; justifying war efforts to significant neutrals; destroying an enemy's morale; improving one's own morale (both among troops and on the home front); inducing an enemy to surrender; assisting with the conduct of a war (e.g. obtaining permission to search a house, gaining the support of local leaders, soliciting information, keeping civilians from battle zones, countering rumors, and assisting with relief efforts), and perhaps most important, resolving a conflict and preventing war in the first place or, having failed in this regard, creating the basis for reconciliation among combatants so that not just the war but the peace is won as well (see Armistead 2004; Blinken 2002; Crossman 1952; Dougherty 1958; Goldstein and Findley 1996; Radvanyi 1990). History also teaches that these information campaigns can be quite competitive, nasty, and brutish (for two examples, see Bytwerk 2005; Sarhandi and Boboc 2001). Soft power has a very hard edge.

There are three ways to change minds and behavior:

1   The control of critical resources or what is called *power* (Emerson 1962; Pfeffer and Salancik 1978), including its most extreme application, war. The use of power also includes promises and threats designed to induce or deter behavior (George 1991; Jervis, Lebow, and Stein 1985).
2   The use of outright *deception* to lead others to believe they are doing X when in reality they are doing something else (Dunnigan and Nofi 2001; Holt 2004; Latimer 2001; Schiffman 1997).
3   The use of *social influence*, by which I mean any noncoercive technique, device, procedure, or manipulation that relies on the social-psychological nature of human beings as the means for creating or changing the belief or behavior of a target, regardless of whether this attempt is based on the specific actions of an influence agent or the result of the self-organizing nature of social systems.

In other words, social influence uses tactics that appeal to our human nature to secure compliance, obedience, assistance, and behavior and attitude change. A social influence campaign can consist of propaganda – the use of often short, image-laden messages to play on the target's prejudices and emotions (Pratkanis and Aronson 2001) – as well as other forms of persuasion, such as debate, discussion, argument, and a well-crafted speech.

In this chapter, I explore the nature of the use of social influence in war and international conflicts. I begin by looking at the *centers of gravity* in influence campaigns – pivotal points in an information campaign that are likely to determine the outcome. Next, I describe some of the social influence tactics that have been investigated experimentally by social psychologists. Then I turn my attention to the social-psychological reception of information. Finally, I attempt to begin a discussion of the possibility of conducting influence campaigns in a manner consistent with the values of a democracy.

Centers of gravity in social influence campaigns

One of the central goals of military planning is to attack what Clausewitz (1832/1984) termed the center of gravity of the enemy – "the hub of all power and movement, on which everything depends" (pp. 595–6). Social influence campaigns, like traditional warfare, have certain hubs on which everything depends. In this section, I identify nine centers of gravity for social influence campaigns to aid with the development of a strategic plan of influence. I view these centers of gravity not simply as principles of social influence, but as decisive points – that is, the key points in a conflict where all of one's social influence is brought to bear to produce a favorable and decisive outcome.

Center of gravity 1: the primacy of strategic attack

"Thus, what is of supreme importance in war is to attack the enemy's strategy" (Sun Tzu *c.*500 BCE/1963, p. 77). The strategy of a social influence campaign consists of goals (which attitudes, beliefs, and actions are to be created or modified), targets (those who are to be influenced), and means and operations (the messages, tactics, media to be used and how a goal will be reached). In a social influence campaign, just as in physical warfare, the influence strategy of adversaries and competitors must be attacked. For example, in World War I, Woodrow Wilson's rally cry, "The war to make the world safe for democracy," trumped Germany's coalition of convenience and naked aggression. (However, when Wilson's idealist vision did not become a reality, his rally cry served to strengthen postwar cynicism and isolationism; see Lerner 1949). In World War II, the Allied information campaign did not directly attack Hitler's propaganda of pan-Germanic unity, nor did it provide an overarching theme for the war other than the defeat of Hitler (Carroll 1948; Lerner 1949). Indeed, Joseph Goebbels (1948) worried in his diaries that Allied propaganda would attempt to divide the German public by attacking "Nazism" but not the German people as a whole. Nevertheless, the psychological warfare division of Norfolk House did attack Nazi strategy at the operational and tactical level, such as the removal of Nazi troops from Rome by undermining the Nazi image as a defender of the faith; countering Nazi propaganda themes; and strengthening the psychological impact of the Allied invasion at Normandy (Carroll 1948). During the Cold War, Radio Free Europe took the strategy of undermining totalitarianism by encouraging independent thinking, fostering evolutionary developments that weakened Soviet control, and supporting nationalistic movements (Puddington 2000). The success of an influence strategy is dependent on the state of the current environment and the reactions of competitors, as determined by centers of gravity in an influence campaign.

Corollary 1a

Strategic traps: An effective social influence campaign is one that traps and constrains enemy action and weakens and divides support for the adversary; conversely, an ineffective influence campaign is one that can serve as a trap for one's own cause. In World War I, the Germans used their technological superiority to dominate North Atlantic shipping lanes through U-boat attacks of ships bringing supplies to Britain and Europe. The British neutralized this advantage by calling attention to the savagery of the attacks, especially those on civilians (e.g. the *Lusitania*). The Wellington House influence campaign effectively trapped the Germans into a choice between either continuing the attacks and risking condemnation of world opinion (ultimately bringing the Americans into the war) or abandoning their technological advantage. In World War II, Hitler used the themes of Aryan superiority and Germanic unity to marshal support for early invasions of Austria and other Eastern European countries and to encourage appeasement based on the notion that Hitler sought only to consolidate Germanic groups. Despite earlier victories, these themes were not a strong enough platform to support Hitler's goals of world domination, and they served to motivate opposition in regions without strong Germanic allegiance, thereby requiring him to rely on quislings and raw power to advance and to maintain the fruits of his aggression. Hitler's information campaign became self-trapping and self-defeating. In the Vietnam War, the Vietcong were successful in portraying the Americans as imperialists and as the descendents of French colonialism. Thus, American forces were trapped by this social influence campaign: legitimate efforts to help the Vietnamese were seen as Yankee imperialism and Vietnamese officials sympathetic to America as mere puppets, and the imperialism theme would serve to divide the American people in their support for the war.

Corollary 1b

360° strategy: A robust social influence strategy is one that anticipates the direction of the conflict in terms of changing circumstances and develops a general framework capable of adapting to conditions. A successful influence campaign is one that adopts general themes (e.g. "The war to end war"; "Stop American imperialism") that can be adapted to fit new targets and situations. It provides a platform for launching a campaign of influence. At any given point in time, the significant actors in a conflict can change as the domain of dispute grows and contracts. Coalitions may realign and split; actors may decide it is in their best interest to wage an information campaign against their former allies. A robust strategy anticipates these changes. Over time, objectives and goals may be modified and further developed. As Lerner (1949) pointed out, victory might change the objectives, such that winning the peace (and not just the war) comes to the fore.

Corollary 1c

Control of information strategy: Strategic influence must be offensive and not reactive. Wallace Carroll (1948), director of the US Office of War Information in London during World War II, observed that for the first two years of the war the Allied information campaign was mostly defensive, countering Axis propaganda to the effect that Britain and Russia would not survive and that the United States would be ineffective in its war efforts. At best, such a campaign is merely a holding pattern in a stalemate of words, but in most cases the use of tactical influence for defensive purposes is the first sign of defeat. Typically, unless the enemy makes a strategic mistake, the best that can be hoped for in a defensive campaign is a tie – that is, the adversary does not gain any ground. (Nevertheless, a defensive campaign is superior to leaving enemy propaganda unchecked and uncountered.) The Allied information campaign met with greater success when it went on the offensive by trumpeting the inevitability of Nazi defeat and raising doubts in the minds of the German public about the war effort.

Corollary 1d

Plans and objectives: Social influence campaigns require a long-range plan with objectives. The successful influence campaign is one that develops a long-range plan that integrates and advances overall policy objectives (Carroll 1948; Crossman 1952; Katz, McLaurin, and Abbott 1996; Lerner 1949). At the same time, the influence consequences of state actions (military or political) need to be considered, either as a reason against such action or, if the action is unavoidable, to mitigate the negative impact of the action. As famed journalist and USIA director Edward R. Murrow put it, public diplomacy should be included at "the takeoffs and not just the crash landings" of foreign policy (quoted in Petersen 2003, p. 8). An ideal long-range plan should involve specific communication goals organized around national strategy and objectives, but with local input and feedback to allow quick, effective implementation. Long-range planning also employs a full conflict analysis (from looking for ways to resolve disputes in the preconflict period through to the postconflict plan for peace). Without such long-range planning, influence campaigns may do more harm than good. Conflicting messages from the same source cancel out and undermine the legitimacy of the effort (Carroll 1948). If not carefully timed, an influence campaign can create false hopes with negative consequences such as the "V sign for Victory" campaign on continental Europe in World War II (Crossman 1952) and the Hungarian uprising of 1956 (Puddington 2000).

Corollary 1e

Inadequacy of advertising and branding: In competitive influence campaigns between nations or other groups, niche strategies of the type suggested by branding theory of advertising are ineffective. Consumer marketing often relies on the concept of product positioning or niche: multiple brands can exist in a product category as long as each appeals to a unique market segment. In social influence warfare, ideas compete head-to-head with other ideas for dominance. As Herz (1949) noted in his comparison of information warfare with advertising, "Praising the excellence of our [cause] is not only secondary but beside the point. As we have seen, it would be difficult to sell the beauty [of our cause] to potential customers who are day in, day out told that [our cause is] a danger and a menace" (pp. 476–7; see also Crossman 1952, pp. 324–5).

Center of gravity 2: Morale Beneffectance

The battlefield in a social influence campaign is the morale of combatants, civilians of belligerent states, and neutrals and third parties who may become engaged in the conflict. Morale is the desire and willingness to start and then continue a cause or action. As with other forms of motivation, it can be described by a principle of beneffectance (Greenwald 1980; Greenwald and Pratkanis 1984; see Tugwell 1990 for a similar perspective): Morale is a function of the perceived legitimacy and goodness of the cause or action (beneficence or doing good) and the expectations for success of the cause or action (efficacy or a sense that "I can"). At any given moment, morale is determined by events (e.g. the execution of nurse Edith Cavell, the Dresden bombing, the Tet offensive), public opinion (the beliefs, attitudes, stereotypes, wishes, and prejudices of participants; see Pratkanis 1989; Pratkanis and Greenwald 1989; see also Center of gravity 5), information and propaganda (e.g. the interpretation of or "spin" placed on events), the relationship of the individual to the group (e.g. a sense of shared purpose and equal sacrifice; Watson 1942), and the participants' desire to maintain the self-perception that they are good and effective actors (often accomplished through dissonance reduction; Aronson 1969; Festinger 1957; see Center of gravity 6). Success in a social influence campaign is measured by the reduction of an enemy's morale (or its components) and the strengthening of one's own morale.

To illustrate the role of morale beneffectance in warfare, consider Figure 3.1, a reproduction of a leaflet used by the Allies on German forces during World War II. Major Martin Herz (1949), chief US leaflet writer during the war, describes this leaflet as one of the most effective (after safe conduct leaflets) of the war. Notice how the leaflet undermines expectations of success (e.g. providing details on why German defeat is inevitable) but does not raise defensiveness (see Center of gravity 6) by blaming the inevitable defeat on the German soldier (e.g. it is not the soldier's fault that he lacked proper armament) or by challenging the rightness of the German cause. Instead, the leaflet reinforces the image of the German soldier as a good and effective actor with a new beneficial goal – not of fighting for a doomed cause but of saving his and other's lives – and provides an excuse for surrender (the alibi of material inferiority).

Other common ways of influencing beneffectance include (1) attacking the goodness of a cause through atrocity stories (Read 1941) and demonizing the

ONE
MINUTE

which may save your life

Read the following six points carefully and thoroughly. They may mean for you the difference between life and death.

1. In a battle of material, valour alone cannot offset the inferiority in tanks, planes and artillery.

2. With the breaching of the Atlantic Wall and of the Eastern Front, the decision has fallen; Germany has lost the war.

3. You are not facing barbarians who delight in killing, but soldiers who would spare your life if possible.

4. But we can only spare those who do not force us, by senseless resistance, to use our weapons against them.

5. It is up to you to show us your intention by raising your arms, waving a handkerchief, etc., in an unmistakable manner.

6. Prisoners-of-war are treated decently, in a fair manner, as becomes soldiers who have fought bravely.

Yon must decide for yourself. But, in the event that you should find yourself in a desperate situation, remember what you have read.

Figure 3.1 A translation of ZG 108, one of the most effective Allied leaflets used during World War II. The leaflet was designed to be used in areas of stiff enemy resistance. The reverse side contained specific instructions on how to surrender.

enemy (Keen 1986; White 1991); (2) bolstering the goodness of one's own cause through patriotism (Bar-tal and Staub 1997); (3) undermining the enemy's perceived efficacy by making defeat appear to be inevitable (for example, in Word War II, Allied leaflets reminded German soldiers of their inferior small arms, Allied-sponsored astrologers predicted Nazi defeat, and the Belgian underground ran a "V for Victory" campaign; Harris 1967); and (4) bolstering one's own sense of efficacy by concretely demonstrating the enemy's vulnerabilities and specifying a plan of action for success (Snavely 2002). In shaping morale, it is generally easier to share a hatred than to share an aspiration (Watson 1942, p. 34), especially when that hatred is centered on a perceived injustice (see Center of gravity 7).

Corollary 2a

Expectations: The perceived success or failure of an action is determined in part by whether or not it succeeded or failed to meet expectations. An expectation is a belief about the future. Expectations serve as reference points by which events are judged. For example, after an Allied bombing run, Goebbels often initiated rumors magnifying the number of German civilian deaths. Later, he would issue an "official" report with correct and lower casualty figures to encourage the German public to view the attack as not so bad after all (Quester 1990; for an Allied example, see Crossman 1952, p. 327). In contrast, the Allied invasion at Anzio in World War II was accompanied by a leaflet making specific predictions of success that were not borne out, and German propagandists used these leaflets to undermine Allied credibility (Herz 1949). Quester (1990) points out that the unexpected German air raids on London during World War I (in which 225 tons of bombs were dropped, killing 1,300 persons) had a dramatic negative effect on British morale, whereas the far more devastating air raids of World War II did not produce a drop in morale and may have stiffened resolve as Londoners realized it was not as bad as anticipated; the British public had expected the use of poison gas and a more severe attack. The firebombing of Dresden had a similar effect on German morale. Finally, US opinions about both the Korean and Vietnam wars became less favorable when military action was stalemated and as a result failed to meet the public's expectations for success (Larson 1996; Mueller 1973). In cases where expectations are not met consistently, trust and credibility are undermined and the communicator loses the ability to persuade a target (see Corollary 3b).

Center of gravity 3: Trust

Without trust and credibility, a social influence campaign is impossible. Ralph K. White, a social psychologist and an architect of the US Cold War information campaign, once stated, "The way into the heart of the skeptical neutralist lies not through artifice but through candor" (1952, p. 540). R. H. S. Crossman, chief of British political warfare during World War II, put it this way: "These are the things that really matter in propaganda. For it is a combination of candour, integrity, and sympathy which demoralizes a totalitarian state" (1953, p. 358). In other words, trust and credibility are an important platform for any communications campaign, especially in a democracy. If a communicator's words cannot be depended on, there is little reason to believe that they will be accepted by a target on a regular basis (exceptions to this rule are rumor and innuendo). Crossman (1953) went on to point out that "crude" propaganda, by which he meant unbelievable and untrustworthy messages, does more to raise the morale of the adversary than it does to advance the goals of the communicator. (Think of the effects of Baghdad Bob – Mohammed Saeed al-Sahaf, the Iraqi information minister – on the American public in the early stages of the Iraq war.)

How can trust and credibility be obtained and maintained? One of the great lessons of every psychological warfare and strategic influence campaign is the importance of speaking the truth (see Armistead 2004; Carroll 1948; Crossman 1952, 1953; Dougherty 1958; Harris 1967; Herz 1949; Puddington 2000; White 1952). If a communication is perceived as propaganda and untrue, it loses its effectiveness. At the same time, however, communications that are true, if not plausibly true in the eyes of the beholder, can be viewed as incredible. For example, truthful messages, such as those describing the provision of desired foods to German POWs in World War II (Harris 1967), the mistreatment of their own war dead by the German leadership in World War II (Crossman 1952), and the valiant actions of Turkish troops in Korea (Dougherty 1958), were all rejected by a target audience because the message was not seen as plausible. (Plausibility is defined in terms of the experience of the target; see especially the principles in Centers of gravity 2, 5, 6, and 7).

Trust and credibility are also obtained and maintained through the establishment of social relationships and roles (known as altercasting; see Pratkanis 2000; Pratkanis, in press). For example, an expert is trusted (more than, say, a child) when communicating about technical matters such as the existence of a 10th planet in the solar system because we expect experts to know such things. In contrast, a child may be more effective than an expert in arguing for nuclear disarmament because a child as communicator places the target in the role of "protector" (see Pratkanis and Gliner 2004/2005; for examples of the use of children in promoting war attitudes, see Sarhandi and Boboc 2001). Similarly, in the world of international affairs, a bully and tyrant places others in a role that produces both compliance (through fear) and resentment that may boomerang against the bully at some point. In contrast, Radio Liberty attempted to create trust through friendship by having their broadcasters "engage in friendly talks" with their targets as opposed to "talking down to their listeners from a platform" (see Puddington 2000, pp. 164–5). Radio Liberty established credibility with its audience by implementing what White (1952) called "respect for the target of the communication."

One difference between a social influence analysis and Nye's (2004) soft power approach is that trust is a platform for communication and not an end in and of itself. For Nye, trust and other positive forms of attraction should inherently draw others to the side of America. A social influence analysis sees this as naive in a competitive and conflicted situation.

Corollary 3a

Listen: A trust relationship begins by listening to one's audience. Recently, the Council on Foreign Relations called for an improvement in America's ability to listen to foreign publics (Petersen 2003; see also White 1952). Listening can build trust relationships in at least two ways. First, it is impossible to communicate with others without knowing their thoughts, beliefs, desires, and concerns. During the Cold War, Radio Free Europe and Radio Liberty made a concerted effort to find out how those behind the iron curtain felt about key issues and about the stations' broadcasts (Puddington 2000). In contrast, the Soviet leadership abandoned opinion polling in the 1930s (as a result of Stalin's disdain for negative information) and as a result had no means of assessing the building discontent within their borders (Benn 1989). Second, the demonstration of a desire to listen to others opens the door for future communication: if the United States is willing to listen to others, then they should be willing to listen to the United States.

Corollary 3b

The significance of deeds: Actions must be in line with words. In a social influence campaign, deeds and actions speak louder than words (Carroll 1948). As Secretary of State Dean Acheson once put it, "What is even more important than what we say to the world is how we conduct ourselves, at home and abroad. The force of example and action is the factor which finally determines what our influence is to be" (quoted in White 1952, p. 540). The failure to act consistently with one's words can have devastating consequences on trust and credibility. As just one example, the Southern newspapers during the Civil War tended to paint a glowing picture of the successes of the Confederate army, resulting in disillusionment and a demoralized public when deeds did not live up to the words (Andrews 1966).

Corollary 3c

Admitting a flaw: A defeat can be used to increase credibility (Crossman 1952). The most unbelievable propaganda is one that consistently paints everything in a positive light, regardless of reality. As Carroll put it, "Unfavorable news must be reported, even to an enemy, if credibility is to be achieved" (1948, p. 125). Carroll's observation is consistent with experimental research showing that admitting to a small flaw increases one's overall credibility (Settle and Golden 1974). During World War II, Allied forces were often quick to report negative news: Allied leaflets often described the "failures" of Allied troops, and the BBC consistently reported bad news, sometimes going so far as to correct Nazi reports even if doing so portrayed the Allies in a less favorable manner. Crossman (1952) reported that there was considerable evidence that criticism of Churchill during the war by the British public served to weaken the morale of the German people. The result of admitting these failures can be seen in a diary entry by Goebbels, who concluded that the British gained more in morale when the BBC covered a defeat than the Nazis gained in announcing a victory.

Corollary 3d

Two sides of the mouth: Black propaganda can undermine trust. Black propaganda consists of messages that are made to appear to come from a source other than the communicator. It is typically used to spread disinformation. According to Herz (1949), such communications have an inherent risk: the detection of black propaganda (or even suspicion of its use) will serve to undermine trust and thus jeopardize the acceptance of other messages (see Crossman 1952 for a similar perspective). As Herz stated: "No nation can talk out of two sides of its mouth at the same time: we cannot on the one hand speak nothing but the truth and then, with a changed voice and pretending to be someone else – but quite obviously still ourselves – say things which we don't dare to say straight out" (1949, p. 483). The use of social influence is often governed by social norms. While it is acceptable to many people if lies and deceptions are told on the battlefield, many also believe that lies and other unfair practices are not permissible in other settings; those who violate these norms run the risk of loss of credibility.

Corollary 3e

Propaganda and trust: Emotional propaganda can produce short-term results but damage long-term trust. Social influence that plays on emotions and prejudices can be an effective means of mobilizing behavior in the short term – even if it is through communications based on lies and deception (so long as the untruth seems plausible to the target). However, such propaganda can negatively affect long-term trust if the target discovers the manipulation. For example, during World War I, the Committee on Public Information (the Creel Commission) rallied US support for the war by disseminating atrocity stories of dubious veracity. When this manipulation was uncovered after the war, it created a sense of cynicism among Americans that made it more difficult to believe in the occurrence of actual atrocities in World War II, such as the Holocaust.

Center of gravity 4: Agenda Setting

Control of the topics of discussion impacts public opinion. Each communication media has an agenda – a list of topics that are the focus of discussion. Issues placed on an agenda appear important and serve to define the criteria used in subsequent decisions. As such, agenda setting becomes an important determinant of public opinion.

The function of agenda setting is nicely illustrated in an experiment conducted by Iyengar and Kinder (1987). In their experiment, students watched edited versions of the evening news over a one-week period. The news shows were altered so that some students received a heavy dose of reports on the weakness of US defense capabilities while others watched shows emphasizing pollution concerns and a third group heard about inflation and the economy. At the end of the experiment, the students rated the target issue – the one that received extensive coverage in the shows they watched – as the issue that was most important for the country to address. In addition, the students supported candidates who took strong positions on their target problem.

Some examples of agenda setting in international conflicts include the British targeting of American opinion makers during World War I with prowar materials written by H. G. Wells, Namier, Toynbee, and others as a means of affecting newspaper editorials; the Allies delivering their newspaper *Frontpost* each week to Axis troops during World War II; Soviet disinformation campaigns such as the one blaming the AIDS virus on US scientists; Serbian media tours of NATO bombing sites; and, just after the Afghanistan invasion, Islamic media discussions of the question "Should there be a continuation of war efforts during Ramadan?" as opposed to other issues (e.g. the cruelty of the Taliban).

How is a communication agenda set? In a nutshell, a communicator sets an agenda by consistently and repeatedly "staying on message" and emphasizing constant themes. However, the specifics of agenda setting vary with the communication medium. Each medium – television news, the Internet, the rumor mill, and so on – has its own set of rules, norms, and procedures for determining what topics are discussed. For example, American mass media place a premium on stories that are new and timely, involve conflict or scandal, contain visual information, are capable of being made personal and dramatic, concern strange or unusual happenings, and fit a current media theme, among other characteristics (Pratkanis 1997). News stories that fit those attributes (e.g. a military scandal) are more likely to be covered than those that don't (e.g. the military building a school). Terrorism is often conducted to appeal to Western media (Nacos 1994), and thus it gains influence beyond its real strength. Entertainment can also be used to attract an audience for agenda setting – for example, Hitler's use of radio and films to draw an audience (Doob 1950) and radio station B92's use of rock music in Serbia to marshal opposition to Slobodan Milosevic (Collin 2001).

One tactic for setting a communication agenda is to control (as opposed to participating in) the flow of information through total information dominance. For example, in World War I, the British cut the trans-Atlantic cables, making it difficult for Germany to communicate with other nations. Information dominance can be obtained through electronic warfare, radio jamming, computer attack, and taking over a local radio station (see Metzl 1997 on the use of information control to thwart genocide campaigns). However, such complete control of information is often seen as a violation of the norms of democracy (except in immediate military operations). Agenda setting that violates the norms for a given medium are deemed unfair and seen as censorship or as manipulation and often results in backlash and a rejection of the message.

Center of gravity 5: Attitudinal Selectivity

An individual uses an attitude selectively to make sense of the world and self. According to the sociocognitive model, an attitude (a cognitively stored evaluation of an object) serves two functions (Pratkanis 1989; Pratkanis and Greenwald 1989; Pratkanis and Turner 1994). First, an attitude can be used to make sense of the world. As such, an attitude is used as a heuristic to bias judgments, attributions, expectations, fact identification, and memory in an attitude-consistent manner (e.g. good [or bad] objects are associated with good [or bad] attributes). Second, attitudes are held to maintain self-worth and are expressed to obtain approval from others, one's self, and reference groups. Given that an individual's attitudes are unique (or at least held as part of a unique group membership), they can provide a frame for selective exposure and attention to events, differential interpretation of stimuli, varied responses to communications, and selective social affiliation, among other behaviors.

Corollary 5a

Segmentation: Groups of individuals with similar attitudes can be targeted with similar messages. Individuals with similar attitudes (and other characteristics) often have similar lifestyles, media habits, and perspectives, making it cost-effective to target such segments in an influence campaign. The specific segmentation scheme to be used depends on the overall strategy (selection of segmentation variables) and objectives (e.g. targeting decision makers, a majority of the population, and so on). For example, during World War II, the Allies used political attitudes to demarcate segments of the German population as a means of developing effective messages (Lerner 1949).

Corollary 5b

The wedge: Given that groups of individuals can vary in their opinions, perceptions of events, and need for dissonance reduction, differential influence can be used to drive a wedge into a coalition and divide support for a cause or action. One strategy for using a wedge to reduce support for an adversary is to target segments of the population that differ on key psychological dimensions with specific, divisive messages. For example, during World War II, Goebbels attempted to drive a wedge between groups in enemy countries by fomenting suspicion, distrust, and hatred (Doob 1950). During the Cold War, Radio Free Europe sought to drive a wedge between the citizens of Eastern Europe and Soviet Russia by emphasizing religious differences (e.g. playing banned Christmas carols) and national pride and between Soviet leaders and workers (e.g. stories on the luxurious lifestyles of leaders and reports identifying corrupt and abusive bosses; Puddington 2000). A second wedge strategy is to target what Herz (1949) called the marginal person or potential "waverer" – the person with divided attitudes about the cause. In such cases, appeals are designed to bolster and support attitudes that are consistent with one's objectives while ignoring those that are opposed to one's cause to avoid a defensive reaction (see Center of gravity 6). The use of a wedge may make later unification of a population difficult if not impossible.

Center of gravity 6: Self-Justification

The tension created by dissonant thoughts produces a drive to reduce that tension and may result in a rationalization trap. According to the theory of cognitive dissonance, when a person holds two discrepant thoughts (such as "I am a good and moral person" and "I just killed someone"), that individual looks for ways to resolve the tension created by the discrepancy (Aronson 1969; Festinger 1957). In such cases, a person engages in self-justification and rationalization to reduce the dissonance by doing such things as derogating (e.g. the person deserved to die), bolstering (e.g. looking for ways to prove one's goodness), seeking external justification (e.g. my religion says it is okay), or reframing the action (e.g. it really wasn't murder). Effective propaganda often stimulates needed tension states or finds a means of reducing dissonance in a manner consistent with the goals of the propagandist.

The potential for self-justification has a number of implications for an influence campaign, including the following (see Bandura 1990):

1   The spreading of rumors and disinformation can serve to resolve dissonance. For example, many in the Muslim world do not believe that the 11 September 2001 terrorist attacks were carried out by Arabs (Peterson 2003).
2   Aggression often conflicts with a person's view of him- or herself as a moral agent and thus requires justification. For example, al-Qaeda required a fatwa (the justification of religious leaders) in order to commit the 11 September murders.
3   A direct attack on beliefs, especially on beliefs closely related to the self, can create defensiveness and thus strengthen the morale to fight. Linebarger (1948) reports, for example, that virulent attacks on Hitler and Germans in general served to stiffen the German will to fight.
4   Discrepant and disagreeable information is often ignored or selectively interpreted. One counter to this tendency is to embed such information in entertaining formats.
5   One reaction to a failing course of action is to escalate commitment to that course of action as a means of convincing oneself that the original action was prudent (White 1968).
6   A call for unconditional surrender (and similar outcomes) often stiffens the resolve of adversaries, as they are faced with a self-threatening admission of defeat (Lerner 1949). To end a conflict – both physical and cognitive – it is

often important to give adversaries an easy way out that is consistent with what you want them to do.

Center of gravity 7: The Seeds of Hatred

Propaganda is most readily accepted by those who are under threat and relatively deprived. Relative deprivation occurs when people's expectations about what they are entitled to exceed what they actually obtain (Crosby 1976; Olson, Herman, and Zanna 1986); it is a feeling that one should be getting more out of life than one is. Relative deprivation can be produced by a number of situations, including a decline in economic and social status or unmet rising expectations for improved conditions. Relative deprivation is experienced as a self-threat (e.g. "perhaps I am not good enough to meet expectations"), which leaves people open to propaganda that will resolve the threat by blaming a scapegoat for the problem (e.g. the Nazis blamed the Jews; al-Qaeda blames the United States), providing a simple solution to the threat (e.g. a Nazi dictatorship or a radical Muslim theocracy), and creating a special identity for the person to restore self-esteem (e.g. "anointed of God"). For example, early Nazi supporters tended to be discontented with their declining economic and social status; notable among them were members of the *Freikorps* and those who had lost wages, jobs, and businesses (Abel 1938). In France and Italy in the 1950s, those who voted communist in elections were likely to feel frustrated that their aspirations and expectations were unfulfilled (Cantril 1958). Islamic and other terrorists tend to be underemployed for their level of education (see Sageman's [2004] data) and feel rejected and second-class before joining terrorist groups (Stern 2003). The frustration of relative deprivation serves as a center of gravity to which hate propaganda flows to create the vanguard of extremist groups and movements. Note, however, that absolute deprivation such as abject poverty is marked by diminished hope, learned helplessness, and resignation and does not necessarily lead to militancy.

Corollary 7a

Perceived injustice: The perception of an injustice is one of the strongest motivations for encouraging attacks, including aggression and war. A perception of injustice occurs when there is a discrepancy between one's actual fate and that to which one feels entitled (Lerner 1981). As a core human motivation, people attempt to restore justice, including resorting to violence and aggression, especially when the injustice is perceived as a threat to one's self-worth (Baumeister and Boden 1998). The history of warfare is the history of the perception of injustice, whether the injustice be real, manufactured, or imagined. For example, the Russian attack on the Turks in the Crimean War was justified by the perceived injustice of being refused the right to protect Christian shrines in the Holy Land and Orthodox Christians living in the Ottoman Empire. Austria's attack (encouraged by Germany) on Serbia, which triggered World

War I, was ostensibly motivated by the assassination of Austrian Archduke Franz Ferdinand. Italy's attack on Ethiopia in 1935, which led to the overthrow of Emperor Haile Selassie, was supposedly in response to a border incident in the city of Wal Wal. The Japanese claimed that their attack on Manchuria in 1931 was motivated by treaty violations by the Chinese (see Lasker and Roman 1938). During the Bosnia-Herzegovina war of the early 1990s, the Serbs put it simply in a propaganda poster: "You have victims for enemies" (Sarhandi and Boboc 2001).

Center of gravity 8: Psychological Reactance

Coercive influence creates resentment and rebellion. We have a general tendency to react against coercive influence, whether that coercion is based on the use of power or deception or is the result of other manipulative processes. This tendency, termed psychological reactance, occurs when one perceives that one's freedom of behavior is restricted; it is an aversive tension state that motivates behavior to restore the threatened freedom (Brehm 1966; Brehm and Brehm 1981). Although the exact response to reactance varies with the situation, two common approaches in conflict situations are (1) an oppositional response (boomerang) of attempting to do the reverse of the reactance-arousing social pressure (a common response to censorship) and (2) the direct elimination of the threat to freedom (Herz 1949). The Aztecs learned the meaning of reactance the hard way. They had violently coerced their neighbors into providing resources for the realm, including victims for ritual slaughter. When Hernando Cortés arrived, these neighboring groups saw their opportunity to reestablish freedom (and avenge a few injustices) by joining with Cortés in the overthrow of the Aztecs. Psychological reactance is an important social influence process in the establishment of democracy and throwing off the yoke of oppressive regimes.

In answer to Machiavelli's question, "Is it better for a prince to be loved or feared?" (to which Machiavelli replied, "Feared"), this analysis suggests that it is important to be both and to follow Teddy Roosevelt's edict to "Speak softly and carry a big stick" (as Nye [2004] also urges). Speaking softly (using social influence, negotiation, and diplomacy) allows conflicts to be resolved without invoking the negative consequences of reactance – an advantage that accrues to mature democracies (Mansfield and Snyder 2005). Coercive bargaining tactics tend to result in increasingly coercive bargaining on all sides (Leng and Walker 1982). In cases where coercive tactics such as threats, punishments, deterrence, and military action must be used, it is most effective to couple the coercion with a "carrot" or face-saving mechanism that allows the target the perception of some freedom of choice. Research on coercion and deterrence has found that they are most effective when the effort has limited objectives, the threat is seen as legitimate (i.e. consistent with social norms), the threat is coupled with carrots for desired behavior, cooperating behavior is reciprocated, and the target has a way out (preferably the desired behavior) to save face (George 1991; Freedman 2004; Leng 1993; Leng and Wheeler 1979).

Corollary 8a

There is more advantage to being perceived as a liberator than as a controlling oppressor.

Corollary 8b

Coercion has time-limited effectiveness. Given that coercive techniques result in reactance, any effects of such tactics are limited (Skinner 1953) and are likely to wear off if the influence agent is perceived to have lost the power to deliver the aversive consequences or is removed from the situation.

Center of gravity 9: The Fog of Propaganda

Misperceptions and distortions of information are common in conflicts, including wars. According to Clausewitz (1832/1984), the fog of uncertainty is a fundamental property of military action (p. 140). The same is true of influence campaigns, which can add to the fog of military action. The propaganda used to support a war can be based on the stereotypes, irrational beliefs, and wishful thinking of a targeted group and thus reinforce the original erroneous thinking. Information disseminated during a clash is typically leveled and sharpened to fit the bipolar theme of conflict. The arousal of a battle can result in black-and-white thinking. During conflict, there is an increased need to know and make sense of the world; any information vacuum will be filled and often is filled with rumor and speculation that increase the density of the fog. This differential flow of influence means that belligerents live in different "reality worlds," resulting in naive realism – a sense that one's own construal (perception and understanding) of the world is real and a failure to correct for the subjective nature of one's own interpretation of events (Ross and Ward 1996). As such, the fog of propaganda can increase the chances for an inadvertent war (White 1968) or continue a war needlessly (White 1984) and make it much more difficult to end a war and win a peace (Pillar 1990).

The chief corrective to the misperceptions of naive realism is what White (1984, 1991) terms "realistic empathy" with an antagonist: an attempt to understand how the conflict looks from the other side's point of view (see also Crossman 1952, p. 328). Empathy is an understanding of the thoughts and feelings of others (in contrast to sympathy or identifying with the feelings of others). Realistic empathy can be achieved by asking such questions as: How would I feel in the situation that my enemy now faces? How do the life experiences and history affect my enemy's judgment? How would I interpret my behavior if I were the enemy? Realistic empathy is the first step toward cutting through the fog of propaganda, identifying the common ground for any attempt at conflict resolution (Arrow *et al.* 1995; Deutsch and Coleman 2000; Fisher, Ury, and Patton 1991), and comprehending the enemy's influence strategy (see Center of gravity 1).

Tactics of social influence

The engine behind every influence campaign is the social influence tactic: a device or procedure that makes use of our nature as human beings to change beliefs and behaviors. Recently, I reviewed the experimental literature and identified 107 empirically tested influence tactics (plus eighteen ways to build credibility; Pratkanis, in press; see also Cialdini 2001; Pratkanis and Shadel 2005). Here, I briefly describe six of these tactics that are used in war and international conflict settings as an illustration of a social influence approach; Table 3.1 lists an additional thirteen influence tactics that are used in war and conflict. A knowledge of social influence tactics is imperative for the conduct of an influence campaign, including planning (an assessment of available options and their limitations), development (creation of influence devices for the situation), operations, and profiling of the adversary (what tactics are most likely to be used and how best to counter them).

Fear appeals. A fear appeal is one that creates fear by linking an undesired action (e.g. the spreading of rumors by civilians in wartime) with negative consequences or linking a desired action (e.g. the surrender of an enemy) with the avoidance of a negative outcome. Fear as an emotion creates an avoidance tendency – a desire to shun the danger. As an influence device, fear has proved to be effective in changing attitudes and behavior when the appeal arouses intense fear, when a specific recommendation is offered for overcoming the fear, and when the target believes he or she can act on the recommendation (Leventhal 1970; Maddux and Rogers 1983). In other words, the arousal of fear creates an aversive state that must be escaped. If the message includes specific, feasible recommendations for overcoming the fear, then it will be effective in encouraging the adoption of that course of action. Without a specific, feasible recommendation, the target of the communication may find other ways of dealing with the fear, such as avoidance of the issue and message, resulting in an ineffective appeal. Interestingly, Herz (1949) identified the same principles of fear in the design of leaflets during World War II and appeals to civilian populations.

Foot-in-the-door. In the foot-in-the-door tactic, a target is first asked to perform a small request (which most people do readily) and then is asked to comply with a related and larger request that was the goal all along. For example, Freedman and Fraser (1966) asked suburbanites to put a large ugly sign stating "Drive Carefully" in their yard. Less than 17 percent of the homeowners did so. However, 76 percent of the homeowners agreed to place the sign in their yards if two weeks earlier they had agreed to post in their windows a small, unobtrusive 3-inch sign urging safe driving. The foot-in-the-door tactic works because it creates a commitment to a course of action and the self-perception that one is the type of person to perform such actions. During the Vietnam War, the Vietcong used this tactic to infiltrate a hamlet. They began by first making a small request of the villagers, such as a drink of water, boys to carry messages, or women to prepare bandages. Building on these small commitments, the Vietcong would

Table 3.1 Some additional influence tactics used in war and conflict

---

Tactic and description

Authority
Authority increases obedience and compliance with commands. Example: In both world wars, Allied leaflet writers found that the addition of authority cues (e.g. embossed seals, the signature of the Allied Commander) increased surrender rates for safe conduct passes.

Define and label an issue
How an issue is labeled controls and directs thought that then impacts persuasion. Example: Goebbels labeled dissent in Britain in World War II "the creeping crisis" in an attempt to make it appear to the German public that the British were losing their will to fight.

Door-in-the-face
Asking for a large request (which is refused) and then for a smaller favor. Example: This tactic is typically used in negotiation settings.

Emotional see-saw
Inducing a change in emotions (happy to sad or vice versa) increases compliance. Example: This tactic is often used in interrogation settings.

Expectations
Options are evaluated by comparison with expectations. Example: Goebbels would preannounce to the German public a higher than expected causality rate for a battle; when the actual death toll turned out to be lower, victory would be claimed.

Imagery sells
Imagining the adoption of an advocated course of action increases the probability that that course of action will be adopted. Example: Effective surrender leaflets encourage imagination of the benefits of ceasing the fight.

Norm of reciprocity
Provide a gift or favor to invoke feelings of obligation to reciprocate. Example: Skilled interrogators often do favors for the person they are interviewing.

Repetition
Repeating the same information increases the tendency to believe and to like that information.

Scarcity
Scarce items and information are highly valued. Example: Marking documents "classified" in Operation Bodyguard led Nazi officials to view them as important.

Social consensus
If others agree, it must be the right thing to do. Example: Mao's propaganda posters often showed many people engaged in a state-approved behavior.

Stealing thunder
Revealing potentially damaging information before it can be stated by an opponent is a counterinfluence tactic to lessen the damage done by an adversary's use of negative information.

Storytelling
A plausible story serves to guide thought and determines the credibility of information.

Vivid appeal
Vivid (concrete and graphic) images can be compelling. Examples: Images such as a Chinese child in the rubble from a Japanese bomb, a child burned by napalm in Vietnam, and prisoners being abused at Abu Ghraib.

then make larger requests, such as support for their war efforts and acceptance of their propaganda (White 1971).

Granfallooning. According to the novelist Kurt Vonnegut, a granfalloon is "a proud and meaningless association of human beings," such as Hoosiers, Buck-eyes, devotees of Klee, or Nazis, that takes on great meaning for those involved. Once an individual accepts a social identity, social influence follows in at least two ways. First, the social identity provides a simple rule to tell the individual what to believe: "I am a _____ [fill in the blank with an identity] and we do and believe _____ [fill in the blank with identity-related behavior and belief]." Second, in specialized cases, some identities become important as a source of self-esteem and locate a person in a system of social statuses. In such cases, influence is based on a desire to stay in the good graces of a positive group and avoid the pain of associating with a derogated identity (Turner and Pratkanis 1998). Some examples of granfalloons include the Nazi brownshirts, the cult of Juche in North Korea, terrorist identities, and bin Laden's use of the Muslim faith. The British were able to stop a communist insurgency in Malaya with appeals to the majority's Muslim identity.

Projection. One way to cover one's misdeed is to employ the projection tactic – accusing another of the negative traits and behaviors that one possesses and exhibits with the goal of deflecting attention away from one's own misdeeds and toward the accused. Rucker and Pratkanis (2001) conducted four experiments in which students were informed that a misdeed had been committed (e.g. lying about intentions, invading another country, or cheating on a test); in the experimental treatments, one of the protagonists in the story accused another of the misdeed. Projection was found to be effective in increasing the blame placed on the target of projection and decreasing the perceived culpability of the accuser. In addition, the effects of projection persisted despite attempts to raise suspicions about the motives of the accuser and providing evidence that the accuser was guilty of the deeds. The projection tactic is a stock technique of authoritarian regimes. For example, in the 1930s, Mussolini accused Ethiopia of provoking Italy as Mussolini invaded the country. After invading South Korea in June 1950, Pyongyang's press and radio claimed that the armies of "the traitor Syngman Rhee" of South Korea had attacked first and that the North Koreans were merely acting in self-defense. During the Korean War, North Korea claimed that US and UN forces were using chemical warfare that caused illness among North Koreans, when in fact the illness was the result of typhus brought to Korea by Chinese soldiers. The use of the projection tactic is inconsistent with the requirement for truthfulness in a democracy.

Damn it, refute it, damn it, replace it. This countertactic for responding to innuendo and disinformation was developed during World War II as a means of rumor control. The tactic entails beginning and ending any refutation of an adversary's disinformation with a clear message that the information is false and negated (damn it). Don't repeat the false information in a memorable manner. The refutation should be logical, brief, factual, consistent, conclusive, and presented calmly. If possible, replace the false information with positive information about the target of the innuendo or otherwise change the topic of conversation.

Jigsawing. Jigsawing is a means of promoting positive relationships among potentially adversarial (ethnic) groups or granfalloons (Aronson *et al.* 1978). The goal is to create mutual interdependence using the equal status contact principle, in which group members possess equal status, seek common goals, are cooperatively dependent on each other, and interact with the support of authorities. This can be accomplished by establishing a situation where people from different granfalloons are required to work together to reach a goal. For example, US aid to war-torn Europe in 1945 (over $9 billion) was originally given piecemeal to previously warring countries and thus served to intensify competition among the factions. Under the Marshall Plan, aid was distributed instead in a manner that required collaborative processes (cooperative dependence), resource sharing (equal status), and joint planning (shared goals) among the various factions (Hogan 1987). The goal was to create an integrated economy and a "United States of Europe" that transcended warring sovereignties.

Social-psychological reception of information: three routes to influence

One of the major determinants of reactions to a persuasive communication or an influence attempt is the target's cognitive responses to the message as they attempt to relate the information to existing attitudes, feelings, knowledge, beliefs, and so on (Greenwald 1968). Indeed, the effective communication can be described by a law of cognitive response: The successful persuasion tactic is one that directs and channels thoughts so that the target thinks in a manner agreeable to the communicator's point of view (Pratkanis and Aronson 2001). However, this brings up a prior question: What determines the thoughts running through a person's head as a message is processed? The answer is in part given by the nature of the influence tactics described in the previous section. Cognitive responses can also be determined by the processing goals of the target, resulting in three routes to influence: peripheral, central, and rationalizing.

Petty and Cacioppo's (1986) elaboration likelihood model is useful for understanding persuasion when the goal of the recipient is to hold a more or less objectively correct opinion about an object or course of action. According to this model, there are two routes to persuasion: peripheral and central. In the peripheral route, the message recipient devotes little attention and effort to processing the message; persuasion is determined by simple cues such as the attractiveness of the communicator, whether or not others agree with the position the communicator is conveying, and the communicator's confidence. In the central route, a message recipient engages in careful and thoughtful consideration of the true merits of the information presented. The person may actively seek out information, argue against the message, and scrutinize the arguments carefully. Persuasion is determined by how well the message stands up to this scrutiny. What determines whether processing occurs via the peripheral or the central route?

Petty and Cacioppo identify two general factors: the motivation to think about the issue (e.g. personally relevant messages are more likely to receive the elaboration of the central route) and the ability to think (e.g. those with cognitive skills are more likely to use the central route).

An experiment by Petty *et al.* (1981) illustrates and validates the elaboration likelihood model. In the experiment, college students heard a message advocating a senior comprehensive exam that all students would need to pass to graduate. For half of the students, the issue was made personally relevant by informing them that the exam was to be adopted next year; the other half were told that it would be adopted in ten years. The students received one of four messages. Half of the messages were attributed to a low-credibility source (a local high school class) and the other half to an expert source (the Carnegie Commission on Higher Education). Petty *et al.* also varied the quality of the arguments in the message, with half of the messages containing weak arguments and the other half containing strong arguments. The results showed that personal relevance determined the route to persuasion. For those students for whom the issue of the exam was personally relevant, the strength of the message was the most important factor in determining persuasion. In contrast, for those students for whom the message was not particularly relevant, the source of the message mattered – the high-credibility source was more effective than the high school class.

The Petty *et al.* research is important for understanding mass media effects in general. Given that much mass media information is not particularly relevant and that the structure of mass media (short, fleeting messages) hinders thought, it is a safe assumption that mass media communication is generally processed in a low-involvement, peripheral route.

In the third rationalizing route to influence, the goal of the target is not necessarily to hold a correct opinion, but to protect and enhance the self. The theory of cognitive dissonance best accounts for this route to influence (Aronson 1969; Festinger 1957; see Center of gravity 6 page 69).

The role of social influence campaigns in a democracy: ethical and legal issues

One consistent point of agreement among those who have practiced the arts of psychological operations (PSYOP) and public diplomacy is that Americans have a strong dislike of and aversion to the use of influence tactics to promote national goals. Elliot Harris (1967), a PSYOP officer during the Korean War, went so far as to title his book on psychological warfare *The "Un-American" Weapon*. At first blush, this aversion appears to be inconsistent with American practice. Our nation was founded on the principle that citizens use social influence and persuasion to make collective decisions as opposed to other devices such as force, divine right, politburo fiat, or the genes of royalty (Pratkanis 1997; Pratkanis and Aronson 2001). We are a nation of sale agents, advertisers, politicians, and lawyers.

Yet, Americans have always had a healthy fear that their government might "PSYOP" them. Crossman (1953) describes a practice during World War II that indirectly justifies these concerns. As part of his PSYOP efforts, Crossman created rumors to undermine enemy morale. His measure of success was whether or not the rumor was later reported to be true by intelligence and news agencies. In retrospect, Crossman doubted "whether it was wise to deceive ourselves so much in an effort to deceive the enemy!" (1953, p. 356). One attempt to resolve this problem was the passage of the Smith–Mundt ban on domestic propaganda, which endeavored to create a firewall to prevent communications by the US government designed for a foreign audience from reaching domestic audiences (Palmer and Carter 2006). However, in an age of cable satellite dishes, the Internet, and near instantaneous worldwide communications, such efforts have become rather ineffectual (see Martin 1958 for a description of failures to control hostile propaganda at the international level).

These legitimate concerns place democracies at a disadvantage. Tyrants, dictators, terrorists, and others who do not value democratic ideals and institutions do not share these concerns about "PSYOPing" citizens. Their overriding goal is to gain power, whether through force, propaganda, or other means. As the Council on Foreign Relations concluded, "The United States needs to be able to counter these vitriolic lies with the truth" (Peterson 2003, p. 36). Indeed, democracies have not just a right but a duty to promote and protect their values. Without a viable public diplomacy effort, responses to tyranny are reduced to two: ignore it (isolationism) or take military action. To generate discussion on this important topic, I conclude this chapter with three suggestions for how to use social influence to promote the national goals of a democracy, along with my own observations on the matter.

Gordon Allport (1942), one of the leading psychologists of his generation and a member of the Society for the Psychological Study of Social Issues' Committee on Morale during World War II, outlined the basis for maintaining morale in a democracy during a war. For Allport, it was acceptable for a democratically elected government to maintain public support and morale (along with influencing other nations) as long as those efforts were consistent with the unique features and values of democracy – voluntary participation, respect for the person, majority rule, freedom of speech, and tolerance, among others. Taking a similar approach, Pratkanis and Turner (1996) present structural characteristics for promoting the deliberative persuasion of a democracy (see also Pratkanis 1997, 2001). The Allport approach has a number of important advantages, including the requirement that a democracy act consistently with its principles (see Corollary 3b page 65) and the use of democratic institutions as a check and balance on the conduct of an influence campaign. It falls short in at least three regards. First, it is vague on operational details. For example, is censorship consistent with democratic values if freely chosen (as was the case, for the most part, in World War II)? Second, authoritarian regimes are not constrained in their selection of influence tactics and thus may gain an advantage. However, to engage in the undemocratic influence tactics of authoritarian regimes is an admission of the defeat of the ways of democracy. Finally, the Allport approach places a premium on maintaining the institutions of democracy as the primary safeguard against the illegitimate use of influence tactics by a government, thus requiring ever-vigilant citizens who are intolerant of intolerance. Such vigilance is often difficult to maintain in times of conflict.

White (1971) saw the problem with the use of influence tactics as a failure to distinguish morally questionable from morally acceptable techniques, which creates a double problem. On the one hand, without a clear understanding of what is a dirty, underhanded technique, the propagandist can use certain tactics without the constraint that might be imposed by conscience. On the other hand, a persuader may be inhibited from doing a proper job by feeling that "use of influence tactics of any kind is wrong." White's morally acceptable influence tactics include getting and keeping attention, getting and keeping rapport, building credibility, appealing to strong motives and emotions, and action involvement (e.g. the foot-in-the-door tactic). Techniques he deems as off limits in a democracy include lying, innuendo, presenting opinion as fact, deliberate omission, and implied obviousness (assuming naive realism). White's approach can be seen as a good first step to legitimizing fair, moral social influence (see Cialdini 1996 for a similar approach). Nevertheless, it runs into problems in implementation. The morality of many influence tactics often depends on the context (e.g. it is generally permissible to use deception on the battlefield or to save a Jew from the Nazis). Some may question whether the use of fear and other intense emotions is acceptable. White would counter that if the object of the fear is real (as opposed to constructed), then it would be immoral not to warn people. But this raises the question of what is a real versus a constructed fear – a question that is often difficult to answer in the fog of war. Finally, as with Allport, White would rule out the use of certain techniques by a democracy that may be employed by authoritarian regimes.

Lt. Gen. William Odom (US Army, Ret.) (1990) presents another approach to justifying the use of social influence in strategic arenas. Odom agrees with other commentators that Americans have a difficult time accepting the concepts of psychological operations, political warfare, and the like and argues that we should abandon such terms altogether. Instead, America should concentrate its political efforts on what Odom sees as the primary foreign policy goal of the United States: to promote democracy where law is primary over even majority parties and serves to protect the rights of the minority. Odom's approach is one of transparency: to be open and up-front about America's goal being to persuade others of the value of democratic institutions. He does not get into the operational details of what are and are not permissible devices, although it can be assumed that his definition of democracy (which includes primacy of law and protection of minority rights) would serve as a check and balance on public diplomacy in manner similar to what Allport had in mind.

Given the comparison of the three approaches above, along with an understanding of the centers of gravity in an influence campaign, we can begin to draw out the characteristics of a public diplomacy organization that both satisfies the need to respond to authoritarian propaganda and allays Americans' fears about the use of influence campaigns. These characteristics include (1) transparency of operations and organization (Americans should be able to see and understand the nature of the influence campaign); (2) reliance on a series of checks and balances to prevent the illegitimate use of influence tactics by the government; (3) consistency of influence campaigns with democratic values, given that actions speak louder than words (see Corollary 3b page 65); and (4) trust on the part of the American people for the public diplomacy organization and consensus on the goals and methods of the influence campaign. The democratic legitimacy of any given influence tactic, like its morality, can vary with the context. However, there is one ethical precept that is invariant across situations. In contrast to the end-justifies-the-means morality of Hitler, Goebbels, and other autocrats, in reality the means determine the end. If an influence agent uses deceit, trickery, or other means of persuasion that are incompatible with democratic values, the effort is likely to produce resentment, lack of trust, reactance, and ultimately more damage than good for the cause. As the Chinese philosopher Mencius warned, immediate goals may be accomplished with devices that produce long-term evil.

The use of social influence in war and international conflict reveals a surprising consistency of technique employed by dictators, demagogues, tyrants, terrorists, and the like. A science of social influence provides an understanding of how to check and counter these tactics and a means for promoting democratic values. With such global developments as the increasing proliferation of weapons of mass destruction and interconnection of the world community, it is imperative that we use this science to create the institutions needed for a more peaceful world.

Bibliography

Abel, T. (1938). *Why Hitler Came to Power*. New York: Prentice-Hall.

Allport, G. W. (1942). "The nature of democratic morale," in G. Watson (ed.), *Civilian Morale* (pp. 3–18). New York: Houghton-Mifflin.

Andrews, J. C. (1966). "The confederate press and public morale," *Journal of Southern History*, 32: 445–65.

Arquilla, J. and Ronfeldt, D. (1999). *The Emergence of Noopolitik: Toward an American information strategy*. Santa Monica, CA: Rand.

Armistead, L. (2004). *Information Operations*. Washington, DC: Brassey's.

Aronson, E. (1969). "The theory of cognitive dissonance: A current perspective," in L. Berkowitz (ed.), *Advances in Experimental Social Psychology*, vol. 4 (pp. 1–34). New York: Academic Press.

Aronson, E., Blaney, N., Stephan, C., Sikes, J., and Snapp, M. (1978). *The Jigsaw Classroom*. Beverly Hills, CA: Sage.

Arrow, K., Mnookin, R. H., Ross, L., Tversky, A., and Wilson, R. (eds) (1995). *Barriers to Conflict Resolution*. New York: Norton.

Bandura, A. (1990). "Mechanisms of moral disengagement," in W. Reich (ed.), *Origins of Terrorism: Psychologies, ideologies, theologies, states of mind* (pp. 161–91). Cambridge, UK: Cambridge University Press.

Bar-tal, D. and Staub, E. (1997). *Patriotism*. Chicago: Nelson-Hall.

Bytwerk, R. (2005). *Paper War: Nazi propaganda in one battle, on a single day, Cassino, Italy, May 11, 1944*. West New York, NJ: Mark Batty.

Baumeister, R. F. and Boden, J. M. (1998). "Aggression and the self: High self-esteem, low self-control, and ego threat," in R. G. Geen and E. Donnerstein (eds), *Human Aggression* (pp. 111–37). San Diego: Academic Press.

Benn, D. W. (1989). *Persuasion and Soviet Politics*. New York: Blackwell.

Blinken, A. J. (2002). "Winning the war of ideas," *Washington Quarterly*, 25(2): 101–14.

Brehm, J. W. (1966). *A Theory of Psychological Reactance*. New York: Academic Press.

Brehm, S. S. and Brehm, J. W. (1981). *Psychological Reactance*. New York: Academic Press.

Cantril, H. (1958). *The Politics of Despair*. New York: Basic Books.

Carroll, W. (1948). *Persuade or Perish*. Boston: Houghton-Mifflin.

von Clausewitz, C. (1832/1984). *On War*, M. Howard and P. Paret, eds and trans. Princeton, NJ: Princeton University Press.

Cialdini, R. B. (1996). "Social influence and the triple tumor structure of organizational dishonesty," in D. M. Messick and A. E. Tenbrunsel (eds), *Codes of Conduct* (pp. 44–58). New York: Russell Sage.

Cialdini, R. B. (2001). *Influence*. Boston: Allyn & Bacon.

Collin, M. (2001). *Guerrilla Radio*. New York: Thunder's Mouth.

Crosby, F. J. (1976). "A model of egoistical relative deprivation," *Psychological Review*, 83: 85–113.

Crossman, R. H. S. (1952). "Psychological warfare," *Journal of the Royal United Service Institution*, 97: 319–32.

Crossman, R. H. S. (1953). "Psychological warfare," *Journal of the Royal United Service Institution*, 98: 351–61.

Deutsch, M. and Coleman, P. T. (eds) (2000). *The Handbook of Conflict Resolution*. San Francisco: Jossey-Bass.

Doob, L. W. (1950). "Goebbels' principles of propaganda," *Public Opinion Quarterly*, 14: 419–42.

Dougherty, W. E. (1958). *A Psychological Warfare Casebook*. Baltimore: Johns Hopkins University Press.

Dunnigan, J. F. and Nofi, A. A. (2001). *Victory and Deceit*. San Jose, CA: Writers Club.

Emerson, R. M. (1962). "Power–dependence relations," *American Sociological Review*, 27: 31–41.

Festinger, L. (1957). *A Theory of Cognitive Dissonance*. Stanford: Stanford University Press.

Fisher, R., Ury, W. and Patton, B. (1991). *Getting to Yes*. New York: Penguin.

Freedman, L. (2004). *Deterrence*. Cambridge, MA: Polity Press.

Freedman, J. and Fraser, S. (1966). "Compliance without pressure: The foot-in-the-door technique," *Journal of Personality and Social Psychology*, 4: 195–202.

George, A. (1991). *Forceful Persuasion*. Washington, DC: Institute of Peace Press.

Goebbels, J. (1948). *The Goebbels Diaries*, L. P. Lochner, trans. New York: Doubleday.

Goldstein, F. L. and Findley, B. F. (1996). *Psychological Operations*. Maxwell Air Force Base, AL: Air University Press.

Greenwald, A. G. (1968). "Cognitive learning, cognitive response in persuasion, and attitude change," in A. G. Greenwald, T. C. Brock, and T. M. Ostrom (eds), *Psychological Foundations of Attitudes* (pp. 147–70). New York: Academic Press.

Greenwald, A. G. (1980). "The totalitarian ego: Fabrication and revision of personal history," *American Psychologist*, 35: 603–18.

Greenwald, A. G., Pratkanis, A. R. (1984). "The self," in R. S. Wyer and T. K. Srull (eds), *The Handbook of Social Cognition*, vol. 3 (pp. 129–78). Hillsdale, NJ: Erlbaum.

Harris, E. (1967). *The "Un-American" Weapon: Psychological warfare.* New York: M. W. Lads Publishing.

Herz, M. F. (1949). "Some psychological lessons from leaflet propaganda in World War II," *Public Opinion Quarterly*, 13: 471–86.

Hogan, M. J. (1987). *The Marshall Plan.* Cambridge, UK: Cambridge University Press.

Holt, T. (2004). *The Deceivers.* New York: Scribners.

Holtman, R. B. (1950). *Napoleonic Propaganda.* Baton Rouge: Louisiana State University Press.

Iyengar, S. and Kinder, D. R. (1987). *News That Matters.* Chicago: University of Chicago Press.

Jervis, R., Lebow, R. N., and Stein, J. G. (1985). *Psychology and Deterrence.* Baltimore: Johns Hopkins University Press.

Katz, P. P., McLaurin, R. D., and Abbott, P. S. (1996). "A critical analysis of US PSYOP," in F. L. Goldstein and B. F. Findley (eds), *Psychological Operations* (pp. 121–48). Maxwell Air Force Base, AL: Air University Press.

Keen, S. (1986). *Faces of the Enemy.* New York: Harper & Row.

Larson, E. V. (1996). *Casualties and Consensus: The historical role of casualties in domestic support for US military operations.* Santa Monica, CA: Rand.

Lasker, B. and Roman, A. (1938). *Propaganda from China and Japan.* New York: American Council, Institute of Pacific Relations.

Latimer, J. (2001). *Deception in War.* Woodstock, NY: Overlook Press.

Leng, R. J. (1993). "Influence techniques among nations," in P. E. Tetlock, J. L. Husbands, R. Jervis, P. C. Stern, and C. Tilly (eds), *Behavior, Society, and International Conflict*, vol. 3 (pp. 71–125). New York: Oxford University Press.

Leng, R. J. and Walker, S. G. (1982). "Comparing two studies of crisis bargaining: Confrontation, coercion, and reciprocity," *Journal of Conflict Resolution*, 26: 571–91.

Leng, R. J. and Wheeler, H. G. (1979). "Influence strategies, success, and war," *Journal of Conflict Resolution*, 23: 655–84.

Lerner, D. (1949). *Sykewar.* New York: George W. Stewart.

Lerner, M. J. (1981). "The justice motive in human relations: Some thoughts on what we know and need to know about justice," in M. J. Lerner and S. C. Lerner (eds), *The Justice Motive in Social Behavior* (pp. 11–35). New York: Plenum.

Leventhal, H. (1970). "Findings and theory in the study of fear communications," in L. Berkowitz (ed.), *Advances in Experimental Social Psychology*, vol. 5 (pp. 119–86). New York: Academic Press.

Linebarger, P. A. (1948). *Psychological Warfare.* Washington, DC: Infantry Journal Press.

Maddux, J. E. and Rogers, R. W. (1983). "Protection motivation and self-efficacy: A revised theory of fear appeals and attitude change," *Journal of Experimental Social Psychology*, 19: 469–79.

Mansfield, E. D. and Snyder, J. (2005). *Electing to Fight: Why emerging democracies go to war.* Cambridge, MA: MIT Press.

Martin, L. J. (1958). *International Propaganda: Its legal and diplomatic control.* Minneapolis: University of Minnesota Press.

Metzl, J. (1997). "Information intervention: When switching channels isn't enough," *Foreign Affairs*, 76: 15–20.

Mueller, J. E. (1973). *War, Presidents, and Public Opinion.* New York: John Wiley.

Nacos, B. (1994). *Terrorism and the Media*. New York: Columbia University Press.

Nye, J. S. (2004). *Soft Power*. New York: Public Affairs.

Nye, J. S., and Owens, W. A. (1996). "America's information edge," *Foreign Affairs*, 75(2): 20–36.

Odom, W. E. (1990). "Psychological operations and political warfare in long-term US strategic planning," in J. Radvanyi (ed.), *Psychological Operations and Political Warfare in Long-Term Strategic Planning* (pp. 8–18). New York: Praeger.

Olson, J. M., Herman, C. P., and Zanna, M. P. (eds) (1986). *Relative Deprivation and Social Comparison*. Hillsdale, NJ: Erlbaum.

Palmer, A. W. and Carter, E. L. (2006). "The Smith–Mundt Act's ban on domestic propaganda: An analysis of the Cold War statute limiting access to public diplomacy," *Communication Law and Policy*, 11: 1–34.

Peterson, P. G. (ed.) (2003). *Finding America's Voice: A strategy for reinvigorating US public diplomacy*. New York: Council on Foreign Relations.

Petty, R. E. and Cacioppo, J. T. (1986). *Communication and Persuasion: Central and peripheral routes to attitude change*. New York: Springer-Verlag.

Petty, R. E., Cacioppo, J. T., and Goldman, R. (1981). "Personal involvement as a determinant of argument-based persuasion," *Journal of Personality and Social Psychology*, 41: 847–55.

Pfeffer, J. and Salancik, G. R. (1978). *The External Control of Organizations*. New York: Harper & Row.

Pillar, P. R. (1990). "Ending limited war: The psychological dynamics of the termination process," in B. Glad (ed.), *Psychological Dimensions of War* (pp. 252–63). Newbury Park, CA: Sage.

Pratkanis, A. R. (1989). "The cognitive representation of attitudes," in A. R. Pratkanis, S. J. Breckler, and A. G. Greenwald (eds), *Attitude Structure and Function* (pp. 71–98). Hillsdale, NJ: Erlbaum.

Pratkanis, A. R. (1997). "The social psychology of mass communications: An American perspective," in D. F. Halpern and A. Voiskounsky (eds), *States of Mind: American and Post-Soviet perspectives on contemporary issues in psychology* (pp. 126–59). New York: Oxford University Press.

Pratkanis, A. R. (2000). "Altercasting as an influence tactic," in D. J. Terry and M. A. Hogg (eds), *Attitudes, Behavior, and Social Context* (pp. 201–26). Mahwah, NJ: Lawrence Erlbaum.

Pratkanis, A. R. (2001). "Propaganda and deliberative persuasion: The implications of Americanized mass media for emerging and established democracies," in W. Wosinski, R. B. Cialdini, J. Reykowski, and D. W. Barrett (eds), *The Practice of Social Influence in Multiple Cultures* (pp. 259–85). Mahwah, NJ: Lawrence Erlbaum.

Pratkanis, A. R. (in press). "Social influence analysis: An index of tactics," in A. R. Pratkanis (ed.), *The Science of Social Influence: Advances and future progress*. Philadelphia: Psychology Press.

Pratkanis, A. R. and Aronson, E. (2001). *Age of Propaganda: The everyday use and abuse of persuasion*, rev. edn. New York: W. H. Freeman/Holt.

Pratkanis, A. R. and Gliner, M. D. (2004/2005). "And when shall a little child lead them? Evidence for an altercasting theory of source credibility," *Current Psychology*, 23: 279–304.

Pratkanis, A. R. and Greenwald, A. G. (1989). "A socio-cognitive model of attitude structure and function," in L. Berkowitz (ed.), *Advances in Experimental Social Psychology*, vol. 22 (pp. 245–85). New York: Academic Press.

Pratkanis, A. R. and Shadel, D. (2005). *Weapons of Fraud: A source book for fraud fighters*. Seattle: AARP Washington.

Pratkanis, A. R. and Turner, M. E. (1994). "Of what value is a job attitude? A socio-cognitive analysis," *Human Relations*, 47: 1545–76.

Pratkanis, A. R. and Turner, M. E. (1996). "Persuasion and democracy: Strategies for increasing deliberative participation and enacting social change," *Journal of Social Issues*, 52(1): 187–205.

Puddington, A. (2000). *Broadcasting Freedom: The Cold War triumph of Radio Free Europe and Radio Liberty*. Lexington: University of Kentucky Press.

Quester, G. H. (1990). "The psychological effects of bombing on civilian populations: Wars of the past," in B. Glad (ed.), *Psychological Dimensions of War* (pp. 201–14). Newbury Park, CA: Sage.

Radvanyi, J. (ed.) (1990). *Psychological Operations and Political Warfare in Long-Term Strategic Planning*. New York: Praeger.

Read, J. M. (1941). *Atrocity Propaganda, 1914–1919*. New Haven, CT: Yale University Press.

Ross, C. (2002). "Public diplomacy comes of age," *Washington Quarterly*, 25(2): 75–83.

Ross, L. and Ward, A. (1996). "Naive realism in everyday life: Implications for social conflict and misunderstanding," in E. S. Reed, E. Turiel, and T. Brown (eds), *Values and Knowledge* (pp. 103–35). Hillsdale, NJ: Erlbaum.

Rucker, D. D. and Pratkanis, A. R. (2001). "Projection as an interpersonal influence tactic: The effects of the pot calling the kettle black," *Personality and Social Psychology Bulletin*, 27: 1494–507.

Sarhandi, D. and Boboc, A. (2001). *Evil Doesn't Live Here*. New York: Princeton Architectural Press.

Schiffman, N. (1997). *Abracadabra! Secret methods magicians and others use to deceive their audience*. Buffalo, NY: Prometheus Books.

Snavely, C. B. (2002). *Historical Perspectives on Developing and Maintaining Home-front Morale for the War on Terrorism*, unpublished master's thesis, Naval Postgraduate School, Monterey, CA.

Sageman, M. (2004). *Understanding Terror Networks*. Philadelphia: University of Pennsylvania Press.

Settle, R. B. and Golden, L. L. (1974). "Attribution theory and advertiser credibility," *Journal of Marketing Research*, 11: 181–5.

Skinner, B. F. (1953). *Science and Human Behavior*. New York: Free Press.

Stern, J. (2003). *Terror in the Name of God*. New York: HarperCollins.

Sun Tzu (*c.*500 BCE/1963). *The art of war*, S. B. Griffin, trans. Oxford, UK: Oxford University Press.

Tugwell, M. (1990). "Terrorism as a psychological strategy," in J. Radvanyi (ed.), *Psychological Operations and Political Warfare in Long-Term Strategic Planning* (pp. 69–81). New York: Praeger.

Turner, M. E. and Pratkanis, A. R. (1998). "A social identity maintenance theory of groupthink," *Organizational Behavior and Human Decision Processes*, 73: 210–35.

Watson, G. (1942). "Five factors in morale," in G. Watson (ed.), *Civilian Morale* (pp. 30–47). New York: Houghton-Mifflin.

White, R. K. (1952). "The new resistance to international propaganda," *Public Opinion Quarterly*, 16: 539–51.

White, R. K. (1968). *Nobody Wanted War: Misperception in Vietnam and other wars*. New York: Doubleday.

White, R. K. (1971). "Propaganda: Morally questionable and morally unquestionable techniques," *Annals of the American Academy of Political and Social Sciences*, 398: 26–35.

White, R. K. (1984). *Fearful Warriors*. New York: Free Press.

White, R. K. (1991). "Enemy images in the United Nations–Iraq and East–West conflicts," in R. W. Rieber (ed.), *The Psychology of War and Peace: The image of the enemy* (pp. 59–70). New York: Plenum.

4 Jihadi information strategy

Sources, opportunities, and vulnerabilities

Glenn E. Robinson

> We are in the midst of war, and more than half of that struggle takes place on an information battlefield; we are in an information war for the hearts and minds of all Muslims.
>
> Ayman al-Zawahiri[1]

Osama bin Laden and al-Qaeda face the formidable challenge of selling a societal vision for which there is only a small demand throughout the Muslim world. Bin Laden's austere and puritanical ideal society, ruled by authoritarian means, simply has no appeal for the vast majority of Muslims. Yet bin Laden and al-Qaeda have garnered significant popularity in the Muslim world, in the process becoming a strategic headache for the United States and its allies. The relative success they have enjoyed despite the unpopularity of their view of what constitutes a proper Muslim society can be attributed largely to their innovative and nimble information strategy. They have been consistently able to frame issues in a plausible and compelling way, and they have taken full advantage of the many missteps of their adversaries, especially the United States. In spite of these successes, al-Qaeda faces structural-ideological limits to its power: it has little chance of ever becoming more than a strategic gadfly, albeit one that can kill thousands of innocents.

Jihadism is a small subset of Islamism. In every Muslim country today, there is a significant minority that supports an avowedly Islamic approach to governance. These minorities reject the secularist wall between mosque and state and seek to solve governance problems through an unapologetic application of Islamic principles, institutions, and concepts to real-world problems. The vast majority of such Islamists seek to implement their program nonviolently, and they gladly participate in the political system where allowed. Periodically Islamist political parties come to power, such as in Turkey. Elsewhere, they have played important roles in parliamentary politics, such as in Jordan, Morocco, and Kuwait. Although electoral results vary with time and place, in democratic elections Islamist parties typically receive 20 to 25 percent of the popular vote.

Jihadists, by contrast, espouse the use of violence to attain their political objectives. While there are different types of jihadi groups, as I discuss below, they all have in common the purposeful resort to political violence as a necessary means to achieve their goal of creating an Islamic state. The political process is rejected as unjust, corrupt, ineffective, or an otherwise inappropriate means of redressing grievances. Nationalist jihadi groups, such as Hizbullah and Hamas, are able to attract high levels of public support, whereas all other types of jihadi groups typically remain marginal to society, have little or no political infrastructure, and enjoy the active support of only tiny minorities of the population.

In this chapter I focus on jihadi information strategy, and on that of al-Qaeda in particular. I argue that al-Qaeda's success in becoming a real player on the regional and international stages is far more the result of an effective information strategy than of any military capabilities or inherent political support it possesses. Put another way, in the absence of political missteps on the part of the United States and its allies, al-Qaeda would have little chance of being seen as a credible player in the Muslim world. More likely, it would be seen as an anachronistic criminal enterprise with no popular support. The chapter is divided into three sections. In the first section, I trace the ideological innovations that have marked modern jihadi thought, beginning with the grandfather of Arab jihadism, Sayyid Qutb (d. 1966) and concluding with Osama bin Laden (although the latter, it should be noted, has always been much more a successful marketer than an ideological innovator). The second section proposes a fourfold typology of jihadi groups in terms of ideology and targeting that allows the drawing of more nuanced and theoretically informed distinctions between very different types of jihadi groups. The third section concentrates on how al-Qaeda has marketed its message – that is, how it frames current issues in order to take advantage of opportunities and limit vulnerabilities.

Jihadi ideological innovation from Qutb to bin Laden

The formulation of jihadi ideology is of recent origin, dating to the 1960s and the struggle for control over the postcolonial state. The ideological founder of Sunni Arab jihadism was Sayyid Qutb, a diminutive Egyptian educator, prolific writer, and employee of Egypt's ministry of education. He lived in the United States from 1948 to 1951, primarily in Greeley, Colorado. Qutb's time in America seems to have begun the process of his radicalization; he complained about rampant racism against Arabs and Muslims in the United States, the moral degradation of American society, and US support for the Jewish theft of Palestine. On his return to Egypt in 1951, Qutb joined the Muslim Brethren organization, which had recently been declared illegal but was still quite prominent. Qutb and the Muslim Brethren supported the 1952 Free Officers coup that brought Gamal Abd al-Nasir and his cohorts to power. Tensions between Nasir's government and the Muslim Brethren steadily worsened, and in 1954 the Brethren attempted to assassinate Nasir. A crackdown on the organization followed, and Qutb, as one of its leaders, was arrested and imprisoned. He was released a decade later, in 1964, after a personal appeal by President Abd al-Salam Aref of Iraq. Qutb was arrested again in 1965 for calling for the over-throw of the "apostate" Egyptian regime, and a year later he was hanged.

Qutb's long years in prison and the torture he and his fellow prisoners endured completed his radicalization. He came to view the structure of the Nasir state as inherently unjust and illegitimate, and he set about constructing a theory to explain why this was the case. Qutb's theory of jihadism was most starkly laid out in his short book *Ma'alim f'il-Tariq* (Signposts on the Road), the book that ultimately cost him his life. Virtually all of the major conceptualizations and ideological innovations that inform Sunni jihadism can be found in Qutb's work. Later ideologues made further contributions, especially Muhammad Abd al-Salam Farag, another Egyptian, who was hanged for his participation in the assassination of Anwar Sadat in 1981. Nationally oriented jihadist groups, such as Hamas and Hizbullah, contributed their own ideological innovations as well.

Still, it would be fair to say that all Sunni jihadi thought is essentially derived from the ideas of Sayyid Qutb. Qutb was primarily influenced by the South Asian scholar Mawlana Mawdudi, whom he met in Cairo, although the radical-ization of Mawdudi's stream of salafi thought came from Qutb.[2] Qutb's theory of jihadism also owes a great deal to Leninist views of both the state and the proper nature of direct action against the state.

While Shi'i jihadism shares many significant features with its Sunni counter-part, there are also major distinctions. The distinctive Shi'i contribution to the jihadi discourse was Ayatollah Ruholla Khomeini's unique interpretation of *velayat-i faqih*, or the rule of the jurisconsult. Like other ideologues, Khomeini took a culturally authentic institution and reinterpreted its meaning in a way that made it virtually unrecognizable to traditional jurists – but made it a potent political tool in Khomeini's information strategy.

Ideological component parts

Instead of creating new vocabularies, jihadists radically reinterpret existing, and culturally authentic, concepts and institutions. In some cases, they simply make things up and claim them as tradition. In this regard, jihadists are no different from nationalist ideologues, who frequently invent histories, meanings, and symbols but claim them as original and authentic.[3] This reinvention of tradition in the construction of a theory of jihadism has a fairly small number of key com-ponents. Classically trained clerics (*'ulama*) inevitably decry the lack of know-ledge of jihadist ideologues because of their "wrong" interpretation of these old Islamic concepts. The war of words between the 'ulama and jihadi ideologues has often been intense. In the following paragraphs, I discuss some of the most important reconfigured concepts.

Jahiliyya

In traditional Islamic thought, the historical epoch prior to the prophet Muham-mad's introduction of Islam to the world was known collectively as *jahiliyya*, a period of ignorance and barbarism. In the absence of God's final prophet and the revelations given to him that form the Qur'an, full truth and salvation were unknowable. Idolatry and lawlessness were the norm. *Jahiliyya*, therefore, was understood chronologically to refer to the period prior to the seventh century CE.

Sayyid Qutb radically reinterpreted *jahiliyya* by arguing that the term did not merely refer to a chronological period but could also be an attribute or character-istic of a society or state. For Qutb, it could apply to contemporary societies in the West that he believed were characterized by immoral and licentious behav-ior. But it could also refer to regimes in Muslim countries that had turned their backs on Islam. The underlying problem was a form of modern slavery that characterized contemporary secular systems of government. Man-made legisla-tion and political systems would improperly empower one man over another, allowing men to lord over others. Indeed, adhering to such man-made institu-tions, instead of those God commanded, constituted a form of *shirk*, or idol worship.

For Qutb, Muslim regimes that denied Islam by ruling through non-Islamic means were guilty of *jahiliyya*, and their rule was inherently marked by injus-tice. Such regimes (and even states) needed to be replaced by Islamic ones, by force where necessary. Armed jihad was the tool to use in undertaking such action. Nasir's Egypt, as a secular and socialist state, was guilty of *jahiliyya* and all the attendant injustice that flowed from *jahili* rule (as Qutb had personally experienced in prison). It was Qutb's call for starting the process of regime change at home, by deposing Nasir, that led to his execution. Qutb's radical reconceptualization of *jahiliyya* has become the primary tool of jihadist groups everywhere to delegitimize the state and call for its overthrow.

Jihad

Jihad, meaning "struggle," is another old and revered concept in Islam that has been imbued with new meaning by the jihadi movement. In traditional interpre-tation, jihad has two dominant meanings. The "greater jihad" constitutes the per-sonal struggle of all Muslims to resist temptation, to become more righteous individuals, and to build a better Muslim community. The "lesser jihad" is com-monly translated as "holy war" and refers to the protection of the Muslim community through a call to arms.[4]

Jihadists have ideologically transformed the concept of jihad in three import-ant ways. First, they have raised the status of jihad to that of the five traditional pillars of Islam (the testament of faith, prayer, fasting, charity, and pilgrimage). Muhammad Abd al-Salam Farag entitled his jihadist manifesto "Al-Farida al-Gha'iba" or "The Absent Duty," referring to jihad. He argued that jihad – jihad of the sword – is rightfully situated at the very center of Islam but has been downplayed and misinterpreted by the 'ulama, especially in recent centuries. In order for Islam to have its rightful place in the world, the true meaning of jihad needs to be recovered.[5]

Second, Jihadists reject the overly strict interpretation of jihad as a defensive war. Traditional interpretation maintains that violent jihad is called for when the 'umma, or Islamic community, is under threat. By definition, then, jihad is always defensive in nature when it is legitimately employed. While bin Laden is careful in his public statements to adhere to the notion of "defensive jihad" – a reaction to a Crusader–Zionist alliance to attack Islam – other jihadists have argued that an overreliance on the defensive nature of jihad not only is a misinterpretation of the concept but also is at odds with Muslim history. They note that the great expansion of Islam in the seventh and eighth centuries through military conquest, often under the banner of jihad, could hardly be considered defensive.

A third departure from the jihad tradition is the notion that the call to violent jihad is obligatory for every individual Muslim (*fard 'ayn*, a mandatory individual duty). Typically, campaigns of the "lesser jihad" have been community obligations, not individual ones. Community obligations are typically organized by the broader community or the government and involve only those Muslims who are in a position to help. Bin Laden's 1998 fatwa (issued with Ayman al-Zawahiri and three other jihadists) calling for every Muslim to kill Americans and Jews under the banner of Islam was thus unusual. It was also odd for someone who is not a member of the 'ulama to issue a religious opinion.

Interestingly, jihad is supposed to remedy the problems of discord and chaos, or *fitna*, both at the individual and societal levels. Yet, jihadists have used the concept of jihad in a way that has enhanced internal discord within the Muslim community.

Waqf

Waqf is a religious endowment (pl: *awqaf*) given by individuals to the 'ulama for the benefit of their mission and institutions. Waqf has historically been an important source of revenue that has allowed the 'ulama to maintain financial independence from the state. Such an endowment can be given in perpetuity, as grants of land for cemeteries and mosques typically have been, but they have more often been granted for specified periods. For example, an endowment might consist of the profits from a particular agricultural field for, say, twenty years.

The Palestinian nationalist jihadi group Hamas has turned this traditional charitable concept on its head. Hamas argues that all of Palestine is *waqf* given by God for use by Muslims until the Day of Judgment. As such, it cannot be divided and given away to non-Muslim use in whole or in part. Hamas's goal in proclaiming that Palestine is *waqf* was, of course, to create an ideological justification for trying to prevent the Palestine Liberation Organization from cutting a deal with Israel to create a Palestinian state. Hamas's argument is innovative in two ways. First, *waqf* is not endowed by God (at least not historically) but by human beings. To state that God engages in provision of religious endowments is a novel argument. Second, Hamas conflates *waqf* with sovereignty, when in fact *waqf* is a private donation, not a declaration of sovereignty. A Muslim in the United States may provide a religious endowment of land for the benefit of his local Islamic center; that charitable gesture does not convey any change in the status of sovereignty governing the land.[6]

Haj Amin al-Husayni, a Palestinian religio-political leader in the 1920s and 1930s, preceded Hamas in arguing that Palestine is *waqf*, although he was even more creative in his reinterpretation. He argued that since the area around the revered al-Aqsa mosque in Jerusalem is considered protected or consecrated land in Islamic tradition (much as one finds in Mecca and Medina today), it effectively constituted a type of *waqf*. Moreover, Husayni asserted that those protected lands were not limited to Jerusalem but extended to include all of historical Palestine. Husayni's creative approach to *waqf* was largely forgotten until Hamas rediscovered and reconfigured the concept.[7]

While the innovation of *waqf* is unique to Hamas and Palestinian jihadism, it does reflect how jihadi ideologues imbue new meanings into old concepts.

Velayat-i faqih

The examples of ideological innovation discussed thus far come from the world of Sunni jihadism. The same process has also been at work in the world of radical Shi'ism. The most famous example of Shi'i ideological innovation involves the tradition of *velayat-i faqih*, or the province of clerical authority. Historically, this concept referred to the areas of responsibility and expertise of members of the 'ulama. For example, the protection of orphans and widows was a typical responsibility that fell under the rubric of *velayat-i faqih*. Domains of specialized expertise held by members of the clergy also constituted provinces of special authority.

In Islamic tradition, *velayat-i faqih* has never meant that the Muslim clergy should hold ultimate political power in a country, but that is precisely what Ayatollah Khomeini argued in a series of lectures in Najaf, Iraq, in 1970. This radical new interpretation of the meaning of clerical responsibility was – and is – rejected by virtually all top clerics in Shi'i Islam, including Ayatollah Sayyid Ali Husaini Sistani in Iraq. Still, it provided the ideological foundation for the postrevolutionary political order in Iran. Khomeini's clerical superior, Ayatollah Kazem Shariatmadari, was placed under house arrest after Iran's revolution until his death for outspokenly denigrating Khomeini's revolutionary reconceptualization of *velayat-i faqih* as clerical political rule.

Jihadi ideological creativity has not been limited to pouring new wine into old bottles. In some cases, new concepts have been created to address strategic problems. Perhaps the most famous ideological creation has been of the dichotomy between the "near enemy" and the "far enemy." This distinction, first coined by Muhammad Abd al-Salam Farag, has since been popularized by Osama bin Laden. Farag and the vast majority of jihadists argued that the focus of their violence must be the near enemy: their local state. While recognizing the importance of a far enemy, usually the United States, the chances of jihadi success were far better, jihadists believed, when they attacked weaker regimes close at hand, such as in Egypt and Saudi Arabia. Conversely, bin Laden argued that it was futile to only attack local regimes when what was keeping them in power was a far enemy: the United States. According to bin Laden, if the chain linking Riyadh and Washington, or Cairo and Washington, were broken, the near enemy would crumble easily. While bin Laden's argument had a certain logic to it, attacking the United States was a highly unpopular move among his fellow jihadists, who accused him of committing a dangerous and even suicidal move that would set back the cause of jihad for generations.[8]

One final remark on jihadi ideological innovation is in order: modern jihadism is distinctively Leninist. Although for obvious reasons jihadi ideologues do not cite Lenin as an inspiration, their concepts and logic, especially Sayyid Qutb's, betray this influence. Having been educated in Egypt in the 1940s, Qutb would certainly have been exposed to Lenin's writings. Two key concepts from Qutb come straight from Lenin: *jama'a* (vanguard) and *manhaj* (program).

Lenin argued that the spontaneous workers' revolution that Marx anticipated could not happen without the work of a highly educated and dedicated vanguard. The workers themselves were too enmeshed in the capitalist system to undertake revolution. Qutb made precisely the same argument for the Muslim world. The vast majority of Muslims were too caught up in and corrupted by the system of unjust and anti-Islamic rule to know how and when to take up arms against the state. A dedicated vanguard of jihadi cadres was needed to organize direct action against the state. Farag furthered Qutb's argument for the centrality of a vanguard when justifying the assassination of Sadat and the (foiled) coup attempt in 1981. Lenin's insistence on the centrality of the vanguard's having a detailed and coherent program for undertaking and then consolidating the revolution was likewise echoed, with an Islamic tone, in Qutb's writings.

A typology of jihadism

Jihadi information strategies vary with the type of jihadi group. Hamas's message is significantly different from al-Qaeda's, which bears little resemblance to that of the Armed Islamic Group – known as GIA, its French acronym – in Algeria. Drawing on the environmentalist slogan to "think globally, act locally" is a useful way to tease out some of the principal differences between the major jihadi streams. Two variables are at work: ideology (national or transnational) and locus of violence (national or transnational). As Table 4.1 shows, a matrix with these variables provides for four different types of jihadi groups, making distinctions that closely match the empirical evidence.

Traditional jihadi groups

The jihad movement launched by Sayyid Qutb still accounts for most jihadi groups around the world. These groups seek to overthrow the regime of their home country. Both their ideology and the locus of their violent actions focus on

Table 4.1 A typology of jihadi groups

Locus of violence	Focus of ideology	
	National	*Transnational*
National	**Traditional jihadi groups** (e.g. GIA, Gama'a Islamiyya, Syrian Muslim Brethren)	**Al-Qaeda franchise groups** (e.g. Saudi Arabia, Morocco, Indonesia, Iraq, Turkey, Egypt, UK, Spain)
Transnational	**Nationalist jihadi groups (e.g.** Hamas, Hizbullah, Chechnya, Iraq)	**Al-Qaeda**

regime change at home and the establishment of an Islamic state. These groups tend to be marginal and to have a small number of committed cadres, typically numbering in the scores or hundreds.

Egypt has provided the most fertile ground for such jihadi groups. Beginning in 1974, a procession of Egyptian jihadi groups undertook violent attacks against state institutions, including an armed attack on Egypt's military academy, the abduction and murder of the former minister of religious endowments, and the assassination of Anwar Sadat and a concomitant attempted coup d'état in 1981. Egyptian jihadism plateaued in the early 1990s when the al-Gama'a al-Islamiyya (the Egyptian Group) and the Tanzim al-Jihad, along with a broader coalition of Islamists, seriously threatened the survival of the Hosni Mubarak regime (and of Mubarak himself). A ferocious counterinsurgency campaign, along with strategic missteps by the jihadists, effectively defeated the jihadi groups by 1998.

Algeria's GIA fought a terror war against that country's military junta throughout the 1990s as part of a broader Algerian civil war. Tens of thousands of civilians died in this campaign, many in quite brutal and grotesque ways. The GIA fits the category of a traditional jihadi group, focused on the overthrow of its home government, although the group also set off several small bombs in France. These attacks were not designed to open a new transnational front, but rather were meant to deter the French government from too openly supporting the Algerian regime in its attempt to crush the rebellion.

The Muslim Brethren, which operates in numerous Muslim countries, is typically an Islamist group, not a jihadi one, but there have been some exceptions. In Syria during the period of 1976–82, Syria's Muslim Brethren took up arms against the "infidel Alawi" regime of Hafiz al-Asad. The Alawi (Shi'i) sect is disproportionately represented in the top echelons of Syria's Ba'thist government, which rules over a largely Sunni Muslim population. The Muslim Brethren's resort to jihadi violence ended badly for them with the February 1982 leveling of the city of Hama by the Syrian military. Hama was a bastion of Brethren support, and its destruction killed about 10,000 people – and ended the revolt.

Information strategies for traditional jihadi groups typically follow the logic of their ideology. The message is to the point:

The regime and its state institutions are jahili, and they are past the point of reform. The injustice, cronyism, and corruption that you see all around you are direct results of jahiliyya. Only by ridding ourselves of this awful regime can we build a truly Islamic state where justice, founded on the shari'a, will prevail.

Jihadi violence usually follows the logic of this message, focusing on trying to assassinate key regime leaders and attacking important institutions of the state, often those associated with internal coercion.

Traditional jihadi information strategies are undermined when groups resort to unjustified violence. Employing violence outside the parameters of societal tolerance has backfired against jihadi groups and has allowed regimes to gain public support in annihilation of the groups. This happened most notoriously in Egypt and Algeria in the 1990s, when local jihadi groups went on an orgy of violence against innocents. In Egypt, those innocents were often foreign tourists, such as in the infamous massacre of fifty-eight foreigners, many of them Swiss, who were visiting Luxor in 1997. The jihadists sought to undermine the Egyptian tourism industry, which they considered essential to the regime's survival. The strategy backfired, resulting in quiet public support for the regime's draconian measures to eliminate the jihadi threat. In Algeria, the GIA's over-the-top violence, both in terms of the number of innocent civilians they killed and the gruesome methods they used in the butchery, likewise backfired in the struggle to win hearts and minds. Even the unpopular and illegitimate regime (having taken power in a coup) was able to parlay public disgust at the GIA's violence into support for crushing the group.

Unable to successfully market extreme violence to their target audience, traditional jihadi groups have backed away from such tactics in recent years. In the case of the Tanzim al-Jihad in Egypt, all violence was publicly renounced, and the group apologized to its victims.[9]

Nationalist jihadi groups

Many of the most prominent jihadi groups in the world today are nationalist in their orientation and frame their irredentist struggle as one of regaining lost lands from a foreign (usually infidel) occupier. Because they are essentially nationalist groups, these jihadi organizations tend to have significantly higher levels of popular support. The most well known nationalist jihadi groups include Hamas (fighting Israeli occupation), Hizbullah (fighting Israeli occupation and, to a lesser degree, politically fighting against any renewal of Christian dominance in Lebanon), and the Basayev group in Chechnya (fighting Russian occupation).[10] In each of these cases, and especially Chechnya, there are other, smaller nationalist jihadi groups operating alongside their more famous compatriots. In addition, some jihadi elements of the Iraqi resistance to US occupation are likewise nationalist in focus, although the exact nature of the relationship between nationalist jihadis in Iraq and the al-Qaeda franchise led by Abu Musab al-Zarqawi until his death in 2006 is unclear.

Nationalist jihadi groups frame their public messages in a very different manner than traditional jihadists do. *Jahiliyya* figures far less prominently in nationalist jihadi discourse, and instead the unjust occupation of national lands by an infidel power takes center stage. In this regard, calls for jihad fit perfectly in the traditional interpretations of when armed jihad is justified: in defense of the Muslim community and its lands. While nationalist jihadi groups focus their ideology entirely on serving their national group, their loci of violence typically cross national boundaries to target the infidel occupier. But what constitutes a national border is not always clear. Nationalist jihadi groups are more vague in framing their message of where their claimed Muslim lands end and the justly held lands of their infidel opponent begin. This is an internal contradiction without a clear and ideologically justifiable solution. Hamas is internally torn over the extent of their territorial claim: the maximalist camp argues for all of historical Palestine, while the pragmatist camp settles for the 1967 lands (Gaza, West Bank, and East Jerusalem). Hizbullah's general refusal to fight beyond Lebanon's international border with Israel – and thus to implicitly accept Israel's existence – is hard to square on ideological grounds but makes much pragmatic sense.[11] Will Chechen jihadists be satisfied with a Russian departure from Chechnya, even though all of the surrounding lands were once Muslim as well? Like Hizbullah, this is another likely example of an ideological "no" but pragmatic "yes" answer.

Nationalist jihadi groups have an enormous advantage over other jihadi groups in terms of maintaining popular support. While the other three jihadi types must always be cognizant of political blowback caused by excessive violence against innocents, this is not a concern for nationalist jihadists. Chechen jihadists do not lose popular support back home when they murder Russian schoolchildren, nor does Hamas when it targets Israeli discotheques and pizza parlors. Periodically, such groups may have to pay a domestic political price if their violence invites an overzealous response by the enemy, but that price is calculated only in terms of pragmatic consequences, not as a verdict on the group itself. As a result, nationalist jihadi groups like Hamas, Hizbullah, and those in Chechnya simply cannot be exterminated by their foreign enemies. They can only be diminished in importance once a legitimate political solution is found, and then they would normally transform themselves into Islamist political parties.

Transnational jihadist groups (al-Qaeda)

Osama bin Laden and al-Qaeda's message is fully transnational, both in terms of its ideological underpinnings and it loci of violence. In terms of the latter, al-Qaeda has successfully targeted US institutions in far-flung corners of the world, including in the United States, Kenya, Tanzania, and Yemen. Ideologically, bin Laden subscribes to the dichotomy between "near" and "far" enemies first promulgated by Farag. Unlike Farag, however, bin Laden and al-Qaeda stress the need to attack the far enemy. The problem is constructed as one where the far enemy, the United States, has long led a campaign of repression against the Muslim world. The motivation attributed to the United States has varied but includes a thirst for Muslim resources such as oil, doing Israel's bidding, and, at the broadest strategic level, engaging in a preemptive war against the only power on Earth that is capable of defeating the American superpower: Islam. Only by keeping Muslims down can the United States stay strong.

Such an ideological worldview thus links otherwise disparate events through their common repression of Muslims. The repression can be applied directly by infidels, as in Iraq, Palestine, Chechnya, Bosnia, and Kashmir; indirectly through repression of true Muslims by apostate regimes, as in Algeria, Saudi Arabia, and Egypt; and even in the guise of international legitimacy through the United Nations, which is seen as a tool of the United States. Western involvement in the conflicts in East Timor and Sudan was also conducted at the expense of Muslim interests, according to bin Laden.[12]

The core of al-Qaeda's information strategy, then, is to convince Muslims that the many disparate problems they face all have the same root cause: the West in general and the United States (allied with Israel) in particular. The solution to ending the repression of Muslims is twofold: to expel the United States from the Muslim world, primarily by getting American troops out of the region, and to attack the chain that links local regimes to Washington. Without US troops to repress Muslims, and without the external support that keeps apostate regimes in power, the Muslim world would rise to its rightful place of power and dignity. Historical injustices in places such as Palestine and Kashmir would be easily righted.

Al-Qaeda's information strategy focuses relentlessly on US foreign policy issues, which maximizes its appeal across the Muslim world, where US policy is widely despised. While it hammers on these broad strategic issues, al-Qaeda is virtually silent on its vision for what a proper Muslim society ought to look like. This vision would be highly unpopular among the vast majority of Muslims, for reasons discussed in more detail below. Thus, while al-Qaeda's overall ideology is broad, the organization employs its information strategy narrowly in ways to maximize its impact among Muslims for whom a message of "fighting the siege" appeals.

Al-Qaeda franchises

The main al-Qaeda organization of bin Laden and Zawahiri has largely been neutralized by post-9/11 US military action in Afghanistan and by Pakistani security forces. The loss of the Taliban host regime was a significant blow to al-Qaeda, and living in hiding has undermined the ability of its top leaders to plan and execute violent transnational attacks. But al-Qaeda has become much more than an organization. Because of its effective information strategy, it has become a powerful idea. Young Muslim men who have never been to Afghanistan have taken up the cause of fighting against the American aggression toward the Muslim world that bin Laden has depicted. In some circles, it has become trendy to be an al-Qaeda fighter, and rap videos and calls to glorious jihad spur on the faithful. No actual al-Qaeda experience or training is necessary. Ideas for making bombs, surveilling targets, and implementing attacks are readily available on the World Wide Web.

This idea of fighting has been actualized through the creation of al-Qaeda franchise cells throughout the world. In some cases, these cells have a member or two with some training from al-Qaeda in Afghanistan, but that is not a requirement. Other cells are simply formed by like-minded Muslims who have been provided with plenty of examples for action through television and the Internet. Bin Laden and other al-Qaeda leaders have no hand in organizing these cells or in directing their activities. Nor will they necessarily have any knowledge of their existence.

Al-Qaeda franchises accept the globalist ideology of the parent organization – that this is a grand struggle by all Muslims against Western, mostly US, domination. However, the locus of their violence is local, in part, no doubt, because of the sheer logistical difficulty of attacking targets on American soil. These franchises are the jihadist equivalent of the slogan "think globally, act locally." The US invasion of Iraq seems to have spurred the creation of al-Qaeda franchises worldwide. In the Muslim world, franchise operations operate in Saudi Arabia, where various cells have killed over 150 Westerners and Saudis since May 2003 (a similar number of jihadists have been killed); Morocco, where a local cell carried out five simultaneous bombings of Western targets in Casablanca in May 2003 that left forty-five dead; Indonesia, where suicide bombings in 2002, 2003, and 2005 left nearly 300 people dead, mostly Australians; Turkey, where jihadists carried out two sets of attacks within five days of each other in 2003, killing fifty-seven; and Egypt, where suicide bombings in Sinai resorts that cater to Westerners and Israelis occurred in 2004, 2005, and 2006, leaving 124 dead. In each of these cases, the bombings were carried out by local jihadists inspired by al-Qaeda against putative Western targets (typically luxury hotels and restaurants). Few of the perpetrators are thought to have been trained by al-Qaeda in Afghanistan or otherwise to have had direct contact with the parent organization. An element of the current Iraqi insurgency also fits this description, but its relative strength and numbers are not known.

Perhaps more troubling is the presence of al-Qaeda franchises in Muslim diaspora communities in Europe. Members of these cells are typically immigrants who have lived for some time in the host country, in some cases for many years, and some cell members were born and raised in the host country. Such jihadists tend to come from much more socially marginal elements than jihadists in the Muslim world, who often hail from relatively privileged backgrounds. The 2004 train bombings in Madrid that killed 191 people were carried out primarily by expatriate Moroccans, most of whom had lived and worked in Spain for years. The July 2005 suicide bombings in London that killed fifty-two people were perpetrated by three native-born British Muslims and a convert from Jamaica who had lived in Britain most of his life. There is unconfirmed evidence that two of the bombers had contact with al-Qaeda in Afghanistan.

Al-Qaeda franchises appear to engage in information operations far less than do nationalist jihadists or al-Qaeda, both of which have sufficient autonomy to implement an information strategy. Like traditional jihadists, al-Qaeda franchises are vulnerable to police and security measures if they become too public in their pronouncements. Typically, franchise information operations are limited to videotapes that are sometimes prepared prior to the suicide action. For example, the leader of the London bombers, Mohammad Sidique Khan, from Leeds, made a video before the bombings in which he praised bin Laden, Zawahiri, and Zarqawi as modern heroes and fully affiliated himself with the goals of al-Qaeda.

Marketing the message: opportunities and vulnerabilities for al-Qaeda

Like any strategically minded group, al-Qaeda must be cognizant of the sensibilities of its target audience – in this case, the Muslim world. Al-Qaeda has little hope of winning the hearts and minds of non-Muslims, let alone the desire to do so, and it does not even attempt it. However, winning over Muslim public opinion to the justice and importance of its cause is an essential step to succeeding in its political project of expelling the United States and its political influence from the Muslim world. Establishing a viable Islamic state is nearly impossible with the hegemonic presence of the United States throughout the Muslim world.

Al-Qaeda's information strategy is predicated on its calculation of opportunities and vulnerabilities. It is important for al-Qaeda to be able to shape its message in a way that appeals to a broad spectrum of Muslim opinion and to avoid broadcasting elements of its ideology that do not appeal to most Muslims. In this section, I argue that al-Qaeda has relentlessly focused its information strategy on criticizing US foreign policy – a message that sells – and has purposely avoided marketing its puritanical and authoritarian vision of an Islamic state – a message that does not sell well in the Muslim world.

Public opinion surveys were generally not permitted in the Arab world until the 1990s, when local research centers gradually gained permission from regimes to conduct limited polling. Polling in the non-Arab parts of the Muslim world was generally more tolerated, but certainly not extensive. The terrorist attacks of 9/11 greatly accelerated the accumulation of survey data in the Muslim world, as suddenly major international polling companies became involved. Gallup, Pew, Zogby, and others began extensive surveys of Muslim public opinion in what was sometimes superficially thought of as a rush to find out "why they hate us." American university research units (from University of Michigan, Harvard University, and University of Maryland, among others) also tapped into the polling bonanza. Local research units in the Middle East found themselves the recipients of large grants to gear up their own polling capacities.

This frenzy of surveying Muslims since the 9/11 attacks has had the markedly beneficial impact of generating a credible set of databases on what Muslims actually think about issues and how those thoughts vary (or don't vary) over time and geography. The days when Western scholars and pundits could merely assert what Muslims must think about this or that issue are over.

The surveys demonstrate two major areas of opportunity for al-Qaeda. The first, and by far the most important, lies in the consistently high proportions of negative ratings US foreign policy draws from Muslims everywhere. The two most important issues in this topic are Palestine and Iraq. Because these issues resonate so clearly with most Muslims, al-Qaeda shapes its message around them. The second major area of opportunity concerns the duplicity of local regimes, both through their corruption and through their alliance with Crusader–Zionist imperialism.

Opportunity for al-Qaeda: unpopular US policy

US foreign policy toward the Middle East is broadly resented in the Muslim world. What is reasonably perceived as one-sided US support for Israel at the expense of Palestinians provides fertile ground for recruitment by rejectionist groups of various political stripes; al-Qaeda is the most recent among them. The US invasion and occupation of Iraq has only bolstered bin Laden's claim that the United States is at war against the Muslim world. Tactical missteps on the part of the United States in pursuit of its broader policy objectives – such as Abu Ghraib and Guantánamo – provided further opportunity for al-Qaeda to conduct an effective strategic information campaign.

The Palestine issue presents al-Qaeda its greatest opportunity for gaining recruits and bolstering public support. Palestine consistently appears to be the barometer issue by which Arabs and even most Muslims judge the intentions and credibility of the United States. Surveys document both the centrality of this issue to most Muslims and the degree to which they disapprove of US policy toward Palestine. The first international poll of Muslims that directly tried to assess the importance of the Palestine issue was conducted by Zogby International in 2002. This poll had the advantage of being conducted well before the US invasion of Iraq, so the numbers are not influenced by this major regional event. The poll asked respondents if they had a favorable view of US policy toward the Palestinians.[13] Table 4.2 presents the results.

Table 4.2 Views of US policy toward Palestinians, 2002 survey (percentages)

	Favorable	*Unfavorable*
Egypt	3	89
Saudi Arabia	5	90
Kuwait	2	94
Lebanon	6	89
United Arab Emirates	10	83
Pakistan	10	79
Indonesia	5	78

Source: Zogby International.

A follow-up question asked respondents about how important the issue of Palestine was to them. After all, someone could disapprove of US policy toward Palestinians but really not think it a very important issue. The results, presented in Table 4.3, did not surprise regional experts.

The poll concluded this line of questioning by asking respondents if they would have a different opinion of the United States if it were to apply pressure to create an independent Palestinian state. As Table 4.4 shows, large majorities of respondents indicated that they would have a favorable impression of the United States under those circumstances. This response suggests a vulnerability for al-Qaeda in its information strategy.

The US invasion of Iraq in 2003 was highly unpopular throughout the Middle East – and, indeed, the world – and it caused a sharp drop in favorable impressions of the United States on various issues. For example, polls commissioned by the Arab American Institute in 2004 and 2005 of six Arab countries showed that Arab views of the United States had declined significantly since the invasion.[14] The downward trend from already low ratings was driven primarily by the Iraq war. The decline was well captured in a question asking, "In the past year, how has your attitude towards the United States changed? Is it now ... ?" The responses are summarized in Table 4.5.

Table 4.3 Importance of the Palestinian issue, 2002 survey (percentages)

	Most important, very important	*Somewhat important*	*Not important*
Egypt	80	15	4
Saudi Arabia	64	16	17
Kuwait	76	17	4
Lebanon	78	13	8
United Arab Emirates	64	20	9
Pakistan	82	11	3
Indonesia	65	20	4

Source: Zogby International.

Table 4.4 View of United States if it were to apply pressure to create an independent Palestinian state, 2002 survey (percentages)

	Favorable	*Unfavorable*
Egypt	69	20
Saudi Arabia	79	19
Kuwait	87	6
Lebanon	50	36
United Arab Emirates	67	20
Pakistan	73	13
Indonesia	60	29

Source: Zogby International.

Table 4.5 Change in attitude toward the United States, 2004–5 survey (percentages)

	Egypt	Jordan	Lebanon	Morocco	Saudi Arabia	United Arab Emirates
Better	5	13	21	6	8	8
Worse	84	62	49	72	82	58
Same	11	18	27	21	9	31

Source: Arab American Institute.

Table 4.6 Favorable views of the United States, 2002–6 surveys (percentages)

	2002	2003	2004	2005	2006
Indonesia	61	15	–	38	30
Egypt	–	–	–	–	30
Pakistan	10	13	21	23	27
Jordan	25	1	5	21	15
Turkey	30	15	30	23	12

Source: Pew Global Attitudes Project.

The Pew Global Attitudes Project public opinion surveys have produced similar findings, namely, that Muslim views of the United States are poor and generally getting worse and that the Palestine issue is the driving force behind these negative views.[15] Pew reports annually on favorability ratings of the United States throughout the world. The results consistently show smaller proportions of people in Muslim countries reporting favorable views of the United States since the Iraq invasion. The favorability ratings from Muslim countries in the Pew surveys from 2002 to 2006 are summarized in Table 4.6.

In the 2006 poll, Pew found that significant percentages of respondents in these Muslim countries felt that the Israeli–Palestinian conflict was a "great danger to regional stability and world peace." About two-thirds of Egyptians (68 percent) and Jordanians (67 percent) rated the Palestine conflict as a "great danger" to world peace, as did large minorities in Indonesia (33 percent), Pakistan (22 percent), and Turkey (42 percent).

In short, survey after survey has shown that the Palestine issue is of central concern to Muslims around the world and that Muslims see US policy related to the conflict in overwhelmingly negative terms. This US vulnerability has constituted a significant opportunity for bin Laden's information strategy. Bin Laden focuses constantly on the theme of Palestine in his public pronouncements, knowing that it plays well to the average Muslim.

Indeed, between 1996 and 2001, bin Laden learned to sharpen his message by concentrating on Palestine as a central grievance rather than on the ills of the Saudi royal family. It is not true, as some have stated, that bin Laden never cared about Palestine before he discovered its marketing potential. For example, bin

Laden issued a public statement in 1994 that harshly criticized the "betrayal of Palestine" contained in the 1993 Oslo Accord. But it is true that Palestine now features far more prominently in his public pronouncements because of its strategic reach.

Bin Laden's first major, detailed public statement of jihad, the 1996 declaration of war against the United States, carried few references to Palestine. Bin Laden did decry the "Jewish occupation of Palestine and their murder and expulsion of Muslims there" and linked Palestine to a broader US-led assault on Islam. Still, the clear focus in this long and tedious manifesto was the abuse of Islamic law by the Saudi royal family and the "blatant imperial arrogance" of the US military "occupation" of the Arabian Peninsula that enabled the Saudi regime to rule without accountability. Indeed, bin Laden lamented that "the greatest disaster to befall Muslims since the death of the Prophet is the occupation of Saudi Arabia, the cornerstone of the Islamic world."[16]

Within two years, bin Laden learned to shorten his message and to focus not on the minutiae of Saudi politics but on key hot-button issues for all Muslims. Bin Laden's famous 1998 fatwa fit entirely on one page and spoke directly to the "Crusader–Jewish" conspiracy that had brought misery and humiliation to Palestine, Iraq, and Saudi Arabia. At the time, Iraq was subject to debilitating sanctions that were highly unpopular in the general Muslim community, and the United States was enforcing the sanctions from its military bases in Saudi Arabia and elsewhere. The fatwa called on all Muslims to kill Americans – military and civilian – and their allies. The goal, according to bin Laden, was to "liberate the al-Aqsa Mosque [Jerusalem] and the Grand Mosque [Mecca] from America's grip" on the path to expelling US forces from "all the territory of Islam."

The evolution of bin Laden's information strategy toward issuing shorter, more focused messages and highlighting the Palestine issue continued after the 9/11 attacks. Bin Laden's first public statement after 9/11, broadcast on al-Jazeera television on 7 October 2001, just as US military strikes began in Afghanistan, was designed to rally the Muslim masses to the defense of Afghanistan and al-Qaeda. Using Palestine as the centerpiece of his message, bin Laden noted an American hypocrisy of silence while "Israeli tanks are going in and wreaking havoc and sin in Palestine – in Jenin, in Ramallah, in Rafah, in Beit Jala." He elevated the issue of Palestine to at least equal importance with his earlier concern with Arabia:

> I have only a few words for America and its people: I swear by God Almighty Who raised the heavens without effort that neither America nor anyone who lives there will enjoy safety until safety becomes a reality for us living in Palestine and before all the infidel armies leave the land of Muhammad [Saudi Arabia].

Bin Laden's information strategy has continued to focus attention on Palestine, an issue of tremendous emotional appeal to most Muslims. Muslims commonly view Palestine as an ongoing humiliation of a land stolen from its rightful inhabitants by European invaders over the past century. The Western betrayal of Palestine with the division of the Ottoman Empire at the end of World War I has figured prominently in numerous public statements. In 2006, for example, bin Laden issued a statement aired on April 23 by al-Jazeera that argued:

> The Palestine question is a manifestation of such injustices when the allied forces of the Crusaders and the Zionists decided to hand over Palestine to the Zionists to establish a state after committing massacres, displaced the indigenous Palestinians and brought Jews from all over the world to settle in Palestine. The ongoing injustice and aggression has still not stopped after nine decades, while all attempts to reclaim our rights and exact justice on the Israeli oppressors were blocked by the leadership of the Crusader and Zionist alliance by using the so-called veto power [at the United Nations]. Such attitudes were also reflected by their rejection of the Hamas movement and its victory in the elections. Their rejection of Hamas has reaffirmed that they are really waging a crusade against Islam.

The US invasion and occupation of Iraq has provided bin Laden with ample rhetorical ammunition to use against America as part of his information strategy. Even before the 2003 invasion, bin Laden regularly used the US sanctions policies against Iraq that followed the 1990–1 Gulf War to delegitimize the American presence in the Middle East. The sanctions and especially the US invasion of Iraq were enormously unpopular among Muslims everywhere, and bin Laden was able to capitalize on Muslim anger and resentment. Bin Laden's October 2002 "letter to Americans" illustrates how he typically frames the Iraq sanctions. The letter, in which bin Laden details his grievances against the United States, was posted on an al-Qaeda Website, both in Arabic and in English translation:

> You [Americans] have starved the Muslims of Iraq, where children die every day. It is a wonder that more than 1.5 million Iraqi children have died as a result of your sanctions, and that you have not shown concern. Yet when 3,000 of your people died, the entire world rises up and has not yet sat down.[17]

The US occupation of Iraq has become a central feature of bin Laden's information strategy since April 2003. Bin Laden has framed the invasion and occupation as a logical part of a general war against Islam and, to a lesser degree, as another manifestation of the United States doing the dirty work of that "petty Jewish state," Israel. Both of these frames resonate within the Muslim world. For example, on the eve of the invasion, bin Laden issued an audiotape, aired by al-Jazeera:

> We are following with intense interest and concern the Crusaders' preparations for war to occupy one of Islam's former capitals, loot Muslims' riches

[i.e. oil], and install a stooge government to follow its masters in Washington and Tel Aviv, like the other treacherous puppet Arab governments, to pave the way for the establishment of Greater Israel.... [This is] a corrupt, unjust war that the infidels of America are waging with their agents and allies.[18]

In every subsequent public statement bin Laden made, he hammered home the point of American calumny in occupying Iraq and praised the insurgents who sought to expel US forces. The notion that the United States is leading a broad effort to subvert the Muslim world, either for its own purposes or on behalf of Israel (bin Laden goes back and forth), is widely believed in the Muslim world, allowing bin Laden to tap into an existing resentment. For example, in the 2002 Zogby poll mentioned earlier, Muslims broadly rejected general US policy toward the Arab world. Asked if they had a generally favorable or unfavorable view of US policy toward Arab nations, respondents answered "unfavorable" by about a 9 to 1 margin (see Table 4.7).

The 2006 Pew poll mentioned earlier also captured the broad distaste for both the United States and, to a lesser degree, the American people in five Muslim countries. Basically, bin Laden is preaching to a ready choir when he accuses the United States of implementing an aggressive and hostile policy toward Muslims.

Opportunity for al-Qaeda: unpopular domestic regimes

In the authoritarian states that populate the Muslim world, pollsters rarely get an opportunity to survey a population about their views on their own regimes. There is significant empirical evidence that many of these regimes, particularly in the Arab world, are not well supported by their populations. There is also some survey data that would support this conclusion as well. For example, in a 2005 Arab American Institute survey, respondents in six Arab countries were asked about high rates of unemployment in their own countries; as Table 4.8 shows, nearly half of all respondents blamed nepotism or poor (government-run) schools as the primary culprits.

Table 4.7 View of US policy toward Arab nations, 2002 survey (percentages)

	Favorable	Unfavorable
Egypt	4	86
Saudi Arabia	8	88
Kuwait	3	88
Lebanon	9	86
United Arab Emirates	13	76
Pakistan	18	75
Indonesia	5	77

Source: Zogby International.

Table 4.8 Causes of unemployment, 2005 survey (percentages)

	Nepotism	*Poor schools*	*Nepotism and schools together*
Egypt	20	10	30
Jordan	32	12	44
Lebanon	22	13	35
Morocco	34	19	53
Saudi	28	29	57
United Arab Emirates	44	17	61

Source: Arab American Institute.

In the 2005 Pew poll cited above, about one-third of respondents in Pakistan (37 percent), Indonesia (31 percent), and Lebanon (30 percent) cited dissatisfaction with the government as the primary reason that Islam's political role was increasing in their own societies. Recent evidence from the United Nations Development Program and World Bank databases on corruption and poor governance suggests that there are abundant reasons for Muslims to be generally dissatisfied with their own regimes.[19]

Bin Laden routinely taps into this dissatisfaction in his public pronouncements. His intent appears to be to make Muslims understand that their regimes are not simply corrupt because of their own inherent flaws but that their corruption stems from their connection to the United States and its allies: break the chain that links these regimes to Washington, and you will free yourselves to create a truly Islamic state. For example, in a widely distributed Internet posting ("Depose the Tyrants!") in December 2004, Bin Laden forcefully called for the overthrow of the Saudi regime because of its corruption and ties to America:

> The Saudi regime has committed very serious acts of disobedience [to God] – worse than the sins and offenses that are contrary to Islam, worse than oppressing slaves, depriving them of their rights and insulting their dignity, intelligence, and feelings, worse than squandering the general wealth of the nation. Millions of people suffer every day from poverty and deprivation, while millions of riyals flow into the bank accounts of the royals who wield executive power. At the same time, public services are being reduced, our lands are being violated, and people are imposing themselves forcibly through business, without compensation. It has got to the point where the regime has gone so far as to be clearly beyond the pale of Islam, allying itself with infidel America and aiding it against Muslims, and making itself equal to God by legislating on what is or is not permissible without consulting God.

Bin Laden did not stop at calling for the overthrow of the Saudi regime. In the same message, he cast a wide net to capitalize on discontent throughout the Middle East:

These oppressive, traitorous ruling families in the region today, who perse-
cute every reform movement and impose upon their peoples policies that
are against their religion and worldly interests, are the very same families
who helped the Crusaders against the Muslims a century ago. And they are
doing this in collaboration with America and its allies. This represents a
continuation of the previous Crusader wars against the Islamic world. The
extent to which the Zionist–Crusader alliance controls the internal policies
of our countries has become all too clear to us. For when it comes to American intervention in internal affairs, where do we start? No appointment of a
king or representative can take place without the agreement of America.

The duplicity of apostate Muslim regimes is a constant refrain in bin Laden's
public messages. In his April 2006 audiotape aired on al-Jazeera, bin Laden
returned to the theme of complicit local regimes:

For their part, the rulers of our region consider the US and Europe as their
friends and allies, while looking at the jihad groups that fight against the
Crusaders in Iraq and Afghanistan as terrorist groups. So how can we reach
understanding with those rulers who deny us the right to defend ourselves
and our religion without carrying arms? The net result of their thinking is
for us to abandon jihad and acquiesce to remaining as their slaves.

US foreign policy and the corruption and duplicity of local regimes are the
major pillars of bin Laden's information strategy, but a number of subthemes
gusset his public statements as well. Two are worth noting. First, bin Laden taps
into a broadly shared Muslim sentiment about Jewish conspiracy and power.
Going well beyond denouncing Israel and its policies, he resorts to a generalized
anti-Semitism in which he evokes Jewish conniving and behind-the-scenes
power brokering. Second, he periodically denounces a perceived cultural domination by the West that is leading to the disappearance of true Muslim culture.
The cultural corruption of Islam becomes almost a lament.

Vulnerability for al-Qaeda: its domestic program

The genius of al-Qaeda's information strategy is to make Muslims forget or
overlook its vision of a proper Islamic state. In public statements, bin Laden,
Zawahiri, and other leaders focus almost exclusively on issues of foreign policy
and corrupt local regimes. Only rarely does a statement slip out that hints at their
vision. What is known of al-Qaeda's domestic program comes from two
sources: statements by bin Laden in the abstract and specific praise of examples
of an Islamic state, such as Afghanistan under the Taliban.

Al-Qaeda's vision of an Islamic state is austere, puritanical, and ultimately
authoritarian. It is not shared by many Muslims – indeed, it is strongly rejected
by the vast majority of Muslims. In terms of real-world examples, bin Laden has
often hailed Afghanistan under the Taliban as something approaching the ideal

type of the Islamic state. Prior to the December 2003 overthrow of the Taliban by US forces, bin Laden made a practice of inviting fellow *mujahideen* (Muslims who take up arms to fight a jihad) to come live in Afghanistan. Over the years, some tens of thousands of jihadists from around the world trained in al-Qaeda camps in Afghanistan, but few chose to remain.

While rumors of sometimes testy relations between al-Qaeda and the Taliban persisted, bin Laden regularly praised the "Islamic Emirate of Afghanistan" under the leadership of Mullah Muhammad Omar, or the "amir al-muminin" (an old Islamic term meaning "commander of the faithful," a title also claimed by the king of Morocco). For example, in an audio statement delivered in April 2001 to a large audience of Muslim activists in Pakistan, bin Laden laid out an argument for the centrality of the Islamic state in Afghanistan within the broader Muslim community:

> You yourselves know that God has ordained for the umma in these difficult times to establish an Islamic state that abides by God's law and raises the banner of His unity, and that is the Islamic Emirate of Afghanistan under the leadership of the Commander of the Faithful Mullah Muhammad Omar, may God keep him. So? it is your duty to call the people to commit to this Emirate and to help it in any way they possibly can, and to stand with it in the confrontation of this torrential current of global unbelief. In order to achieve this we urge that you include in the final statement of the conference a call to help the Islamic Emirate of Afghanistan by all possible means, spiritually, financially and verbally.... On this occasion I assure you and Muslims across the world that I submit to God on the duty of allegiance to him.... For Mullah Omar is the ruler and rightful commander who rules by God's law in this age.

Afghanistan under Taliban rule was widely regarded as regressive and antimodern, even among most Muslims. Its peculiar amalgamation of Islamic law and Pushtunwali tribal codes was unique in Muslim history. While it did win some praise for bringing a measure of law and order to a chaotic Afghanistan, its regressive policies in all manner of issues – technology, consumer products, minorities, women, even the rule of law – brought condemnation from the international community, including from Muslim states. While I have not found surveys of Muslim views toward Taliban rule, significant empirical evidence suggests that Muslims widely viewed Taliban rule as an embarrassment to Islam. The low point of Taliban rule was the regime's deliberate destruction of an ancient Buddhist monument in Bamyan, an act that drew the wrath of Muslim states around the world.

Bin Laden praised the destruction of the Bamyan Buddhas as one of Mullah Omar's "historic Islamic positions that affirm his honesty and steadfastness on the path, for which we admire him." Bin Laden and other al-Qaeda leaders have frequently derogated different minorities, including Buddhists, Christians, and Shi'ites. Bin Laden's greatest vehemence is saved for Jews, a regular theme in his public statements. To give but one example, from a February 2003 statement:

> The Jews are those who slandered the Creator, so how do you think they deal with God's creation? They killed the Prophets and broke their promises. These Jews are masters of usury and leaders of treachery. They will leave you nothing, either in this world or the next. These Jews believe as part of their religion that people are their slaves, and whoever denies their religion deserves to be killed. These are some the characteristics of the Jews, so be aware of them.

Bin Laden's regressive vision for Muslim society is clearly a by-product of his Wahhabi belief system. Wahhabism, an eighteenth-century interpretation of Islam formed in the harsh interior of the Arabian peninsula, is austere and puritanical in its approach to Muslim society. It is deeply unpopular in Muslim societies outside the Arabian Peninsula, but it has made some inroads here and there, largely as a result of financial support for its propagation from the Saudi government.

Even in Wahhabi Saudi Arabia, however, support for the consumer lifestyle of modern American capitalism is strong. In the 2002 Zogby poll cited earlier, respondents were asked for their views of modern American cultural products. The response was overwhelmingly favorable across the Muslim world. Table 4.9 summarizes these views on American science and technology, movies and television, American-made products, and American education.

Polling data show a clear disconnect between bin Laden's vision of a proper Muslim society and the strong embrace of the modern consumer world by large majorities of Muslims around the world. The ability of bin Laden and al-Qaeda to gain the kind of stature they have in spite of this disconnection is a tribute to their flexible and nimble information strategy. In public statements they emphasize things on which most Muslims would agree (antipathy toward US policies in Palestine and Iraq, for example) while deemphasizing things for which they would not get broad Muslim support (their domestic vision). As a result, they have been able to minimize the strategic disadvantages for al-Qaeda while maximizing the advantages. The level of confidence Muslims have in bin Laden remains relatively high. Those reporting a lot of or some confidence in the al-Qaeda leader in the 2005 Pew poll included 60 percent in Jordan, 51 percent in Pakistan, 35 percent in Indonesia, and 26 percent in Morocco. By contrast, in Turkey he received only 7 percent and in Lebanon 2 percent. Bin Laden clearly remains a force, largely as a result of his success in winning the battle for the story.

Vulnerability for al-Qaeda: over-the-top violence

All forms of jihadism except nationalist jihadism have suffered from the perception that they have engaged in wanton violence. The use of over-the-top

Table 4.9 Muslim views on modern American cultural products, 2002 survey (percentages)

	US science and technology		US movies and TV		US products		US education	
	Favorable	Unfavorable	Favorable	Unfavorable	Favorable	Unfavorable	Favorable	Unfavorable
Egypt	78	11	53	40	50	45	68	17
Saudi Arabia	71	26	54	42	53	44	58	35
Kuwait	86	12	54	44	57	39	57	29
Lebanon	82	16	64	35	72	25	81	16
United Arab Emirates	81	14	64	32	68	27	79	13
Pakistan	83	14	64	28	75	18	80	16
Indonesia	83	8	77	20	71	15	79	9

Source: Zogby International.

violence has backfired on jihadism, and jihadists have not been able to formulate a compelling information strategy to justify such violence against innocent civilians. The GIA's violence in Algeria clearly undermined the jihadi appeal to the Algerian general population, just as jihadi violence in Egypt paved the way for popular support for severe measures on the part of the regime to repress the jihadists there. Al-Qaeda franchise violence against civilians in Saudi Arabia, Morocco, Egypt, and Indonesia has also helped to turn popular sentiment sharply against jihadists.

Pew polling between 2002 and 2005 showed sharp declines in support for jihadi violence. The evaporation of support was greatest in countries that experienced this violence firsthand. For example, Moroccan popular support for jihadi violence declined precipitously after bombings there. In March 2004, 40 percent of Moroccans said that violence against civilian targets could sometimes be justified; by 2005, only 13 percent thought so. The proportion who said that violence against civilians could never be justified rose from 38 percent to 79 percent in the same period.

Jihadists of most stripes seem to be rethinking the use of violence to achieve their goals, largely because it has been self-defeating in the information war. As noted earlier, Tanzim al-Jihad in Egypt has renounced violence and apologized to its victims. The Syrian Muslim Brethren has also renounced violence and formally seeks peaceful political change. Broad public outrage over the bombings against Muslims by al-Qaeda franchises in Saudi Arabia seemed to prompt a reevaluation. Since nationalist jihadists are generally rewarded politically for their violence (although within limits), there is no reason to expect such groups to renounce violence until there is a political settlement to their irredentist claims.

Policy implications

This analysis suggests both good news and bad news for American strategists in dealing with al-Qaeda's information strategy. Four specific implications are suggested. First, there are limits to al-Qaeda's information strategy. Its regressive philosophy for Muslim society in an Islamic state simply does not appeal to most Muslims. As a result, al-Qaeda has had to limit its information operations to focus almost exclusively on the injustices of US foreign policy and the complicity of local apostate regimes in America's war of domination. These two broad themes make up over 90 percent of all of bin Laden's public messages. It will be very difficult for al-Qaeda to move beyond these themes as it continues to fight for the hearts and minds of the Muslim public. On the other hand, al-Qaeda has succeeded in staying on theme, so limiting its message to these two broad topics has carried it fairly far.

Second, at a tactical level, there is very little US planners can do to influence Muslim public opinion. It is clear that the United States does not have an image problem in the Muslim world, it has a policy problem. Tactical information operations may have some impact at the margins, but there is no evidence that they have or can have a significant impact on the way Muslims view the United States.

Third, at the strategic level, changes in foreign policy can have a significant impact on diminishing the appeal of al-Qaeda's information strategy and bolstering the standing of the United States in the Muslim world. But strategic-level changes are the hardest to implement. Clearly, a more balanced American policy vis-à-vis Israel and the Palestinians would go a long way toward generating goodwill for the United States in the Muslim world. The recent call to reexamine the place of Israel in US foreign policy-making by prominent scholars Stephen Walt of Harvard and John Mearsheimer of the University of Chicago may help to inch along a gradual strategic change to the benefit of US national interests.[20] A reasonably good outcome in Iraq would also help in rebutting a primary message in al-Qaeda's information strategy, although that is looking less and less likely. Iraq's collapse into all-out civil war would constitute a strategic disaster for the United States in the battle with al-Qaeda for the story. The precipitous decline in favorable views of the United States in the Muslim world as a result of the invasion of Iraq demonstrates how flexible and fluid such views are in the face of significant changes in the strategic environment. Creating more functional and accountable governance in the Muslim world would also help the United States at the strategic level, but this issue is highly complex and one over which the United States has only limited power.

Fourth, the information revolution works to the advantage of nonstate actors. In previous decades, the information realm in Muslim societies was dominated by the state, with censored and state-run television, radio, and printing presses. The ability of nonstate actors to break through this information monopoly was extremely limited. This is no longer the case. Beginning with Ayatollah Khomeini's famous distribution of audiocassettes to foment revolution, to the proliferation of satellite television stations and offshore radio, to video recorders and cell phone cameras that can record events anywhere in the world, to the explosion of Internet connectivity and e-mail in the Muslim world just over the past decade, nonstate actors can now compete with states to get out their version of the story. Al-Qaeda has taken full advantage of the information revolution, and it will likely be able to continue to do so for a long time to come.

Notes

1 Letter to Abu Musab al-Zawahiri, July 2005, found by US forces in Iraq (my translation). While most scholars accept the authenticity of the letter, it is still a matter of some debate.

2 The Arabic term *salafi* is also used to describe Qutb's new version of radical Islam. However, the Salafiyya movement includes both violent and nonviolent strands, so it is not coterminous with jihadism. See Quintan Wiktorowicz, "Anatomy of the Salafi Movement," *Studies in Conflict and Terrorism*, 29: 207–39 (2006).

3 There is a large literature on this theme in studies of nationalism. An excellent first source is Eric Hobsbawm and Terence Ranger, eds, *The Invention of Tradition* (1983; repr., Cambridge, UK: Cambridge University Press, 1992).

4 The classic analysis in English on jihad is Majid Khadduri, *War and Peace in the Law*

of Islam (Baltimore: Johns Hopkins University Press, 1955). A much shorter but quite useful overview of the evolution of jihad can be found in Rudolph Peters, "Jihad," in *The Oxford Encyclopedia of the Modern Islamic World* (New York: Oxford University Press, 1995). For a comparison between jihad and the just war tradition in Christendom, see John Kelsay and James Turner Johnson, eds, *Just War and Jihad: Historical and Theoretical Perspectives on War and Peace in Western and Islamic Traditions* (Westport, CT: Greenwood Press, 1991).

5 For a full translation of and commentary on Farag's manifesto, see Johannes J. G. Jansen, *The Neglected Duty: The Creed of Sadat's Assassins and Islamic Resurgence in the Middle East* (New York: Macmillan, 1986).

6 Interestingly, Hamas's argument is the mirror image of the old Zionist conflation of land purchases in Palestine with sovereign right. Zionists purchased less than 7 percent of the lands of Palestine before 1948 and then argued that this land ownership conferred rights of sovereignty.

7 See Hillel Frisch, "Territorializing a Universal Religion: The Evolution of Nationalist Symbols in Palestinian Fundamentalism," *Canadian Review of Studies in Nationalism*, 21(1–2): 45–55 (1994).

8 For an excellent analysis of the debates within jihadi circles over the far enemy concept, and of jihadism more generally, see Fawaz A. Gerges, *The Far Enemy: Why Jihad Went Global* (New York: Cambridge University Press, 2005).

9 Ayman al-Zawahiri and his followers left Tanzim al-Jihad as a result of this turn toward pacifism and linked up with al-Qaeda instead.

10 Shamil Basayev was killed in July 2006.

11 The July 2006 Hizbullah cross-border raid was a notable exception to this policy and was done both in solidarity with Palestinians in Gaza then under Israeli siege and as a means to free Hizbullah prisoners held by Israel. The deadly campaign that erupted from this raid also served the interests of Syria and Iran, close supporters of Hizbullah.

12 An excellent synopsis of bin Laden's worldview as framed through one of his public statements can be found in an audiotape aired by al-Jazeera television on 23 April 2006.

13 Full results and details of the 2002 poll may be found at www.zogby.com/. The same poll showed strong overall Muslim support for American cultural products and values, contrary to popular mythology in the United States.

14 Complete poll results can be found at: www.arabamericaninstitute.org/.

15 The Pew polls can be found at www.pewglobal.org

16 An abbreviated English version of bin Laden's declaration of jihad can be found in Bruce Lawrence, ed., *Messages to the World: The Statements of Osama Bin Laden* (New York: Verso, 2005), from which these quotations were taken. A much longer, full translation, which captures the tedious flavor of bin Laden's detailed critique of the internal workings of the Saudi regime was made by Muhammad al-Massari, leader of the now-defunct Committee for the Defense of Legitimate Rights.

17 Lawrence, p. 164. Unless otherwise noted, the remaining bin Laden quotes come from Lawrence's excellent annotated volume.

18 Ibid., p. 180. Aired on 11 February 2003.

19 United Nations Development Program, *Arab Human Development Report* (New York: United Nations Press, 2002, 2003, and 2004). The World Bank's global governance and corruption database and analysis can be found at www.worldbank.org/wbi/governance/.

20 John Mearsheimer and Stephen Walt, "The Israel Lobby," *London Review of Books*, 28:6 (23 March 2006).

5 Reorganizing for public diplomacy

Carnes Lord

The lamentable condition of American public diplomacy today is widely acknowledged. A raft of studies and reports over the last half decade or so by a variety of official, semiofficial, and independent bodies have told a broadly similar story of institutional ineffectiveness, lack of strategic direction, and insufficient resources.[1] Recently, these criticisms were acknowledged by President Bush and Secretary of State Condoleezza Rice, who in her Senate confirmation hearing promised to make public diplomacy reform a "top priority." In the aftermath of the 11 September 2001 attacks, the elected leadership of the nation summoned the political will to make the most far-reaching changes in the national security bureaucracy in more than half a century in order to enhance the security of the American homeland in the face of the terrorist threat. A similar national commitment to institutional change seems to have emerged in the area of intelligence. Such a commitment does not exist, unfortunately, in the area of public diplomacy, in spite of the administration's proclaimed concern. There are several key reasons for this. Not only is there no real consensus among practitioners or critics of American public diplomacy as to what needs to be done to fix it, but also the nature of the pathologies afflicting it are themselves not well understood.

Public diplomacy pathologies

What accounts for this situation? It is easy enough to blame the end of the Cold War and the loss of that era's widespread sense that the tides of history were flowing in our direction and that the United States needed to do little to explain itself in a world in which liberalism and democracy were suddenly the norm. In fact, however, the problems facing public diplomacy today have much deeper roots. Even during the Cold War, public diplomacy never fully lived up to the expectations many initially held out for it, and it remained for the most part at the margins of American grand strategy. Public diplomacy has been uniquely disadvantaged as an instrument of American statecraft. It alone among the elements of our national security policy lacks a core institutional base, an established infrastructure of education and training, a stable cadre of personnel, an operational doctrine, and roles and missions that are understood and accepted by national security and political elites, let alone by the general public. Not coincidentally, it lacks high-level political support and congressional interest and tends to be chronically underfunded. Moreover, public diplomacy operates in a uniquely challenging domestic environment, one centrally shaped by the fundamental hostility of the media and much of the general culture in the United States to any government involvement in the management of information.

The pathologies of public diplomacy may be traced to three distinct yet interrelated sets of problems. The first has to do with the definition and scope of the subject; the second, with the aura of illegitimacy that surrounds it; and the third, with the difficulty of organizing the government effectively to carry it out. With respect to the first of these, there is a long history of fluctuating and vague terminology in this field that has tended to subvert rigorous thinking about the subject and greatly complicate the organizational challenges associated with it. Fundamental questions remain unresolved about the basic mission of public diplomacy and the level of involvement a variety of government agencies – emphatically including the Department of Defense – should have in it; this confusion is cause for particular concern in the international environment of protracted warfare we face today. The term "information operations" has recently gained traction in the US military as a way to conceptualize and organize a broad and diverse array of "nonkinetic" military capabilities. As yet, however, efforts to align information operations with public diplomacy in the traditional sense of that term – to say nothing of the traditional public affairs function – have not been noticeably successful.

Although organizational issues relating to public diplomacy cannot be separated entirely from the definitional issues I've just alluded to or from longstanding ideological disputes over its legitimacy,[2] there may nevertheless be some utility in focusing narrowly on the organizational dimension. The key question here is whether or to what extent the Department of State is the proper home for public diplomacy or whether consideration should be given to establishing a new agency dedicated to public diplomacy along the lines of the old United States Information Agency (USIA), which in much reduced form was merged into the State Department in 1999.

The appointment in late 2005 of former presidential adviser Karen Hughes as under secretary of state for public diplomacy and public affairs underscores the administration's commitment to significant improvement in the department's performance in this area. It remains to be seen what overall impact this appointment will have on policy or institutional change. The presumption, however, would appear to be that the appointment of Ms. Hughes signals a commitment, at least for the foreseeable future, to maintaining the State Department as the primary locus of public diplomacy within the US government. While such a commitment is by no means indefensible, it would be unfortunate if it were permitted to choke off a potentially valuable and arguably necessary debate over the proper institutional framework for carrying out public diplomacy in the current security environment. In what follows, I attempt to frame such a debate by making the case for reconstituting USIA.

Organizing the information sector

Since the outbreak of World War II, there has probably been more instability in the information sector of the US national security bureaucracy than in any other. During the war itself, long-running and bitter disputes over roles and missions marked relations among the three organizations that had responsibilities in this area – the Office of War Information, the Office of Strategic Services (forerunner of the CIA), and the psychological warfare branch of the army.[3] A residual information service was kept after the war as a semiautonomous bureau of the State Department, but the Voice of America came close to being abolished. With the outbreak of the Korean War, interest in public diplomacy revived within the Truman administration and in Congress, but there was little consensus on how it should be organized. In 1953, after much internal study, USIA was created by President Eisenhower, but largely because of the reluctance of Secretary of State John Foster Dulles to have his department undertake what he saw as an operational or programmatic function. The Voice of America joined the new agency, but the education and cultural affairs function was, rather illogically, retained by State. Meanwhile, the newly formed CIA had set out to create its own broadcasting empire with the creation of Radio Free Europe (1950) and Radio Liberty (1953) as "surrogate" radio stations targeting Eastern Europe and the Soviet Union, respectively, and in addition sponsored an array of political action campaigns directed against the growing threat of Soviet Communism.[4]

The involvement in public diplomacy of a variety of very different and potentially competing organizations raised the question of coordination, and hence in particular the question of the White House role in public diplomacy. As a Defense Department official lamented at the time, "Our psychological operating agencies are like bodies of troops without a commander and staff. Not having been told what to do or where to go, but too dynamic to stand still, the troops have marched in all directions."[5] During the Korean War, the Truman administration created a Psychological Strategy Board under the auspices of the recently established National Security Council (NSC) to organize and spearhead this effort, but it was able to make only limited headway and was abolished early in the Eisenhower administration. The reorganized NSC system of the Eisenhower years included an Operations Coordinating Board, which was assigned public diplomacy and related tools as part of its responsibility for interagency coordination of foreign and national security policy generally, but its public diplomacy focus seems gradually to have dissipated. It was not until the Reagan administration that an effort was made to reestablish an interagency coordinating mechanism for public diplomacy under NSC auspices – again with only limited success. The problem remains a fundamental one that must be addressed in any rethinking of the public diplomacy architecture of the US government.

The 1970s again saw major organizational change. After the exposure of the CIA connection to Radio Free Europe and Radio Liberty on the floor of the US Senate in 1971 and the resulting threat to the radio stations' continued existence, Congress created a new oversight mechanism, the presidentially appointed

Board for International Broadcasting (BIB), in 1973. In 1976, the two radio sta-
tions and their corporate boards were merged into a single entity, "RFE-RL."
Next, attention turned to USIA.[6] In 1977, in the wake of a number of outside
studies of the organization of the foreign affairs agencies of the government,[7] the
Carter administration merged the State Department's Bureau of Education and
Cultural Affairs into USIA and changed the name of the parent agency to the US
International Communications Agency.

The Stanton report and the demise of USIA

In the most influential of the studies just mentioned, the report of the so-called
Stanton Commission,[8] several recommendations were made that, although not
adopted at the time, would shape the course of the debate on these issues in
important ways. Going back to fundamentals, the Stanton Commission identified
the core missions of public diplomacy as follows: exchange of persons, general
information, policy information, and policy advice. (By "policy advice," it
meant the shaping of policy decisions by public diplomacy considerations, as
advocated famously by USIA director Edward R. Murrow.) The commission
then pinpointed as the key problem "the assignment, to an agency separate from
and independent of the State Department, of the task of interpreting US foreign
policy and advising in its formulation." It also noted "the ambiguous positioning
of the Voice of America at the crossroads of journalism and diplomacy." This
analysis led the commission to the findings that (1) USIA should be abolished
and replaced with a new, quasi-independent Information and Cultural Affairs
Agency, which would "combine the cultural and 'general information' programs" of both USIA and State's ECA [Education and Cultural Affairs]
bureau;" (2) State should establish a new "Office of Policy Information, headed
by a deputy undersecretary, to administer all programs which articulate and
explain US foreign policy"; and (3) the Voice of America should be set up as an
independent federal agency under its own board of overseers.

The Stanton Commission recommendations responded to two very different, if
somewhat contradictory, requirements. The first was a perceived need to enhance
the "credibility" of American public diplomacy by increasing its distance from the
government (an acknowledgment of the legitimacy deficit of official information
programs noted earlier). The second was to fix what the commission saw as an
artificial and ultimately dysfunctional organizational separation between the State
Department, which was the agency responsible for formulating US policy, and
USIA insofar as it was responsible for defending US policy and disseminating it
abroad. In a sense, this reform, too, was intended to enhance the credibility of US
overseas information programs, but it would do so by narrowing their distance
from government policy, not increasing it. Henceforth, the State Department
would speak for itself – and thereby necessarily assume greater responsibility for
the accuracy, timeliness, and effectiveness of its information efforts.

As it turned out, the next stage of public diplomacy reorganization in the
mid-1990s, culminating in USIA's merger into the State Department in 1999,

largely embraced the Stanton Commission approach. Although the commission's vision for USIA itself was not realized, the information functions of the old USIA were indeed folded into State, and Congress created the new position of under secretary for public diplomacy and public affairs in the State Department to give public diplomacy new visibility and clout there. At the same time, significant steps were being taken toward creating autonomy for US international broadcasters. Under the International Broadcasting Act of 1994, a new oversight board, the Broadcasting Board of Governors (BBG), replaced the Board for International Broadcasting and was assigned authority not only over RFE-RL and the more recently created surrogate radio stations (Radio Martíand Radio Free Asia) but also over the Voice of America itself. Although the BBG initially operated within USIA and the secretary of state was made an ex officio member, its authorizing legislation gave it virtual operational control of all of US international broadcasting. The stage was now set for the eventual liberation of all US overseas broadcasting from US government control.

The problem with State

There is a great deal to be said in favor of the view taken by the Stanton Commission that the same agency that is responsible for the formulation of US foreign policy should also be responsible for making that policy generally known and for explaining and defending it. Diplomats can make excellent propagandists, and substantive policy experts will always have more credibility than those seen as mere flacks. Moreover, tightening organizational links between policymakers and public diplomatists would seem to result in many administrative efficiencies. It was also eminently plausible to suppose that the State Department, once handed this mission of the former USIA as well as many of the people and resources that had been committed to it, would embrace these responsibilities, support them appropriately, and carry them out effectively. Unfortunately, the reality has been something else entirely.

Allowance, of course, must be made for the inevitable confusion and inefficiencies that attend any major reorganization. Yet it is now some seven years since the State–USIA merger, and there is still no real evidence that the State Department has either the vision or the will to conduct effective public diplomacy. Fundamental problems remain at the level of organizational structure, culture, and leadership.

The State Department's public diplomacy structure after the merger centers on the bureaus of International Information Programs and Educational and Cultural Affairs – both formerly part of USIA – under the direction of the under secretary for public diplomacy and public affairs. In addition, public diplomacy officers in modest numbers are seeded throughout State's regional bureaus, the core policy-formulating element of the department, and are assigned to US embassies overseas. Another important component of the former USIA, the Office of Media Research and Analysis, is attached to the Department of State's bureau of Intelligence and Research. The key problem is that the under secretary has no authority over the public diplomacy functions or personnel not in his or her direct chain of command and has little say over resource issues relating to public diplomacy. Personnel matters affecting public diplomacy officers in the field are handled by the ambassador, not at the Washington level, as was the case with USIA in the past. In addition, senior public diplomacy officers in the regional bureaus are too few and at too low a level (office directors rather than deputy assistant secretaries) to ensure real integration with policymaking, and they have tended to become isolated.

But a more fundamental problem is that of organizational culture. As one public diplomacy officer remarked of the merger, USIA people "have come from an organization that sent out information and arrived in an organization that draws information in and by nature keeps it locked in."[9] The information function has always lacked prestige within the culture of the Foreign Service, and it is currently ghettoized – that is, public diplomacy is a fifth career "cone" within the Foreign Service, distinct from the prestigious political cone. This has meant consistent understaffing and underfunding of public diplomacy activities. State provides inadequate training in the discipline for public diplomacy officers, let alone for its other personnel. Moreover, the highly bureaucratic character of the State Department, with its elaborate and time-consuming system of clearances and paperwork, is antithetical to the requirements of effective public diplomacy, which must be sensitive to news cycles and capable of responding immediately in crisis situations. USIA, by contrast, was programmatically oriented, capable of producing tangible products quickly and efficiently. Finally, with few exceptions, the Foreign Service tends to lack an appreciation of the real nature of public diplomacy and, in fact, to confuse it with public affairs. It sees public diplomacy as fundamentally reactive rather than proactive. Its attitude toward releasing information is to "fire and forget," not to run extended information campaigns carefully calculated to produce certain effects on certain audiences. And it favors blandness and banality over argumentation, controversy, and color.[10]

Another problem, finally, is that the State Department has failed to provide the leadership so clearly needed to overcome these deficiencies. Under the Clinton administration, little high-level attention was paid to any of these issues during the crucial reorganization period, while in the Bush administration, a critical mistake was made in appointing to the position of under secretary for public diplomacy and public affairs an advertising executive without foreign policy expertise or Washington experience – Charlotte Beers, who led the department down the blind alley of "branding" the United States as a Muslim-friendly society. Nor did Secretary of State Colin Powell take any apparent interest in this subject.[11] After Beers's departure in March 2003, the position was left vacant for nine months; her eventual successor, Margaret Tutwiler, departed after only a brief tenure.

It is conceivable that stronger leadership at the political level could effect changes in State Department culture that, over the longer term at least, would enable it to perform the public diplomacy mission satisfactorily. Under Secretary Hughes has in fact demonstrated an understanding of many of the problems just discussed and has taken constructive steps to remedy them. Yet one cannot but be deeply pessimistic about the prospects for fundamental change in State's way of doing business, given the widespread view in Washington today – voiced most forcefully several years ago by former House speaker Newt Gingrich but echoed as well within the ranks of the Foreign Service itself – that the State Department as a whole has not risen to the challenges of the contemporary era.[12] This is a larger issue beyond the scope of the present discussion. But to cite only one problem relevant to our concerns here, the department's failure to create incentives or a career structure encouraging higher education among its officers (comparable, for example, to the uniformed military) has led to acute deficiencies in knowledge of languages, culture, and history throughout the Foreign Service.[13] To be sure, the fault cannot be laid entirely at the feet of the department, as Congress has for many years regularly denied State even minimally adequate funding and personnel levels.

USIA: back to the future?

All this having been said, the obvious question is whether the State–USIA merger was a mistake and ought to be reversed. Critics of American public diplomacy today are divided on this issue. I argue here that the merger was probably an experiment worth trying,[14] but one that has signally failed. It is simply unrealistic to look to the State Department to undertake the full range of public diplomacy activities, to do so aggressively and creatively, and to run "programs" in an efficient and responsive way. Therefore, I argue, USIA will continue to be needed and should be reinvented.[15] At the same time, both on the merits and as a practical matter, it would be wrong to try to undo everything that has been done to improve the integration of public diplomacy and policy at the State Department. On the contrary, I believe it is essential to continue to press State to do more and to do better in this area. And if USIA is to be reconstituted, it is also an open question whether it should be reconstituted in its earlier form.

The Stanton Commission was broadly correct to recommend that State should take over only those USIA functions that directly involve policy information and advice and that USIA should have the cultural function (and the Education and Cultural Affairs bureau). It was wrong, however, in attempting to exclude USIA from any policy-related role, limiting it essentially to educational and cultural affairs and to "general information" about America and American society in a longer-term perspective. The real question is how to sort out the exact division of labor between State and USIA relative to policy-related information – as well as action, an aspect of public diplomacy the commission failed completely to address.

The fallacy of creating a sharp separation between educational and cultural affairs and "policy" is even clearer today than it was in the 1970s. In the context of an international conflict that is not merely ideological in nature but a "clash of civilizations," in Samuel Huntington's now classic phrase, education and culture are front and center in a way they were not during the Cold War. Hence the idea that education and culture represent an arena for essentially nonpolitical interaction with adversaries simply cannot be sustained today, and it becomes difficult to see what advantage is gained from assigning a separate organization to oversee that arena. This is by no means to argue that educational and cultural public diplomacy must in all cases be carried out by a government agency; indeed, there may be great advantages in giving the private sector the lead in many of them. It is only to argue that there is a compelling argument for retaining at least some policy control over these activities.[16]

What State can do

What public diplomacy–relevant functions can be performed only or most effectively out of the State Department? There seem to be two distinct but closely related tasks: ensuring the closest possible connectivity between public diplomacy and diplomacy or policy; and exploiting the synergy between public diplomacy and policy. If this is correct, it suggests a clear standard for judging the sorts of public diplomacy activities that are appropriate for State and those that are not, and at the same time it underlines the importance of fixing some of the organizational problems noted above. Connectivity requires day-to-day collaboration between policy and public diplomacy officers to ensure that the public diplomacy implications of evolving policies are thoroughly understood, that policy itself is shaped by public diplomacy considerations as appropriate, that policy is articulated in a way that takes full account of its public impact abroad, and, finally, that policy is supported in the process of implementation by supplementary information and materials that help explain and defend it (background documents, briefing slides, talking points, "Q&A's," and the like). This kind of day-to-day collaboration clearly has its operational side, but these are not operational activities that can be easily reduced to a "program" to be executed elsewhere; they are (or must be if they are to be effective) inextricably enmeshed in the routine work of the department.

Ensuring this connectivity makes it possible to exploit the synergy between public diplomacy and diplomacy. This is a neglected but nonetheless important dimension of our subject. Diplomacy can support and reinforce public diplomacy campaigns, lending credibility to messages that might otherwise be dismissed by foreign governments or other observers as mere propaganda. And public diplomacy can support diplomacy by bringing pressure to bear on foreign governments through their own media and public opinion. Furthermore, diplomats may engage directly in public diplomacy campaigns, that is, in face-to-face interactions with various nongovernmental foreign audiences; this is an important part of the political action component of public diplomacy. Obviously, all of these things can only occur in and through the State Department.

The organizational implications of all this for the State Department are reasonably clear. Public diplomacy must avoid ghettoization by having a presence in each regional bureau as well as in the functional bureaus. This presence, currently very uneven and in some cases nonexistent, should be at a sufficiently high level to ensure that collaboration actually takes place – that means at the level of deputy assistant secretary (at least in the regional bureaus). The under secretary for public diplomacy and public affairs should be given clear administrative authority over these public diplomacy officers in the regional and functional bureaus (although for operational purposes they would of course answer to their respective assistant secretaries). This is almost certainly the only way to ensure that the integrity of the public diplomacy function is respected throughout the department. It appears that this change is in fact now occurring under the leadership of the new under secretary, Karen Hughes.[17]

At the same time, under this plan the under secretary would lose the two "programmatic" bureaus currently under his or her direct control (the bureaus of Education and Cultural Affairs and of International Information Programs) to a reconstituted USIA, while a new, relatively lean bureau would need to be created – perhaps called simply the Public Diplomacy bureau – to perform functions essential to the core State Department public diplomacy mission. In addition to supporting the under secretary's new administrative responsibilities, this bureau should include a public diplomacy policy planning and resources element and a crisis response element with a small operations center.[18] It would presumably have some regionally oriented structure but should also have sufficient staff to allow flexible task organization to deal with emerging issues and to play effectively in the interagency arena.

With various of its current operational functions stripped away, as suggested here, some may doubt whether there could be sufficient justification for a new Public Diplomacy bureau or, indeed, whether a new entity would be needed at all. Given the enormous daily volume of information and the constant ebbs and flows in US policy, however, it is difficult to imagine that such a staff would not have enough to do. Moreover, an assistant secretary is a valuable asset not only in keeping public diplomacy visible and attended to throughout the department but also in representing State effectively in the interagency arena and abroad. There is plainly much work to be done in our diplomatic dealings with friends and allies to raise awareness of public diplomacy as a strategic tool and to enlist them in multilateral public diplomacy initiatives of all kinds.[19] Such a staff should also be expected to play a key role in maintaining connectivity and coordination with State's very important Public Affairs bureau, which also answers to the under secretary.

Finally, there is the question of the proper locus of a political action component in the State Department and, more generally, the relationship between the responsibilities of the under secretary for public diplomacy and public affairs and State's activities in the area of democratization and human rights. It can certainly be argued that the Bureau of Democracy, Human Rights, and Labor is a more natural fit for the under secretary of public diplomacy than for the under secretary for democracy and global affairs (formerly the under secretary for global affairs), whose span of control currently includes in addition the Bureau of Oceans and International Environmental and Scientific Affairs and the Bureau of Population, Refugees, and Migration.[20] At the same time, a case could be made for enlarging the Democracy bureau to cover a wider range of traditional political action functions. There is already a separate office for International Women's Issues under the global affairs under secretary; this is a subject of great salience for American public diplomacy in the Muslim world and would be an obvious candidate for bringing into closer proximity to State public diplomacy. In addition, new offices could be created for issues relating to youth, religion, sports, and perhaps other such sectors of high significance in international public opinion.

Reinventing USIA

Where, then, does all this leave USIA? The general principle I have been trying to establish is that a revived USIA should no longer have a substantial role in generating and managing day-to-day or routine "policy information," as it did prior to the merger; this function properly belongs with State. But it would retain a number of important roles relating to policy information. Several additional general principles may now be introduced to help define the scope and nature of USIA's activities.

First, USIA should again be seen as *the* institutional base of public diplomacy in the US government. This should be understood to include broad responsibility for training and career development as well as the guardianship of public diplomacy doctrine, something that has been done inadequately if at all in the past. It also implies the reintegration of international broadcasting into the agency. Second, USIA should be the primary operational agency for public diplomacy. Although State's operational role would be recognized and accommodated, USIA would regain responsibility for managing public diplomacy operations in the field, including jurisdiction over the public affairs officers in the embassies.[21] It would also regain ownership of the Media Research and Analysis unit currently in State's bureau of Intelligence and Research as well as the Foreign Press Centers. (By contrast, it would make sense to have the Washington File, a heavily used daily compilation of government-generated information distributed electronically to US embassies around the world, remain at the State Department.) Moreover, USIA rather than State would have the lead role in developing and fielding new technological capabilities in support of public diplomacy – perhaps in a new and closer relationship with the Department of Defense.

USIA's role under this scheme should not be seen as only operational or programmatic. It would have the lead responsibility for research and analysis relating to the foreign media and foreign public opinion and, along with the State Department, would engage in strategic planning relating to public diplomacy campaigns and programs. Like State, it would generate policy-relevant information, but its activities would be more proactive and creative. Its products might range from formal white papers, to primers on American foreign policy or the American political system geared to a variety of foreign audiences, to the production of documentary films or TV features on policy issues such as global warming or on contemporary history. A major function (though one that would need to be shared to some degree with State, Defense, and the intelligence community) would be the monitoring, analysis, exposure, and countering of adversary propaganda and disinformation activities – an occasional USIA function in the past, mostly in wartime, but now arguably required on a sustained basis. As it has done in the past, USIA should administer speakers programs abroad addressing policy issues, and its field officers should engage continuously on these issues with foreign journalists, academics, and other opinion makers. It should attempt to revive its former network of American libraries oversees and oversee major new programs to disseminate books and written materials of all kinds to key foreign audiences. Finally, USIA should facilitate contacts between high-ranking American officials throughout the government and key influentials and audiences abroad, something that has never been done systematically in the past.

This last point is an important one, as it underlines one of the great advantages of having a USIA separate from the State Department. USIA should be seen as serving US foreign policy in the largest sense, not simply American diplomacy; it should work cooperatively with every agency of the US government that has significant overseas presence or interests – the Agency for International Development as well as the departments of Energy, Commerce, Agriculture, Justice, and (particularly) Defense. Again, the USIA envisioned here would retain much of the programmatic orientation of the old USIA. Where it could improve on its predecessor organization is in having greater agility and flexibility, a higher level of creative and analytic capability, and greater engagement in the substance of the issues with which public diplomacy deals. In these respects, it would hark back in significant ways to an older model, the Office of War Information of World War II vintage, which was staffed by a distinguished group of journalists and intellectuals.[22] To build such an organization, however, would require quite a different approach to personnel recruitment and retention than that of the old USIA or of today's State Department. While it will always be desirable to staff such an agency with some number of career civil service and Foreign Service personnel, ways need to be found as well to bring in talented noncareer personnel for limited tours. These could be new PhD's in a variety of relevant fields; they could be midcareer businessmen with significant foreign experience and language skills; or they could be early retirees, particularly former military officers. Some could perhaps be hired on a contract basis rather than as regular government employees. Preserving flexibility in firing as well as hiring should be a consideration in any case, given the high potential for burnout in many public diplomacy jobs. For whatever reason, the two organizations in the US government today whose employees are on average the oldest are the Voice of America and the International Information Programs bureau of the State Department.[23] While it is certainly important to value experience in this field and to acknowledge the importance of husbanding certain esoteric skill sets, it has to be recognized that the current situation is simply not optimal if we are to be serious about improving our public diplomacy capabilities.

Wider issues

Some final remarks may be made concerning the wider institutional setting of American public diplomacy. The idea that a public diplomacy "czar" operating out of the White House can solve all the problems we face in this area is perennially appealing but profoundly misguided. What is needed at the national level is rather two things. First, an effort needs to be made in and through the National Security Council staff to enhance the integration of information and policy at every level of the interagency process. Second, a new entity should be created within the Executive Office of the President with a primarily programmatic focus and an expansive mandate to coordinate public diplomacy operations with international assistance programs and other instruments of American "soft power."

As for the Department of Defense, an adequate discussion is impossible here, but the basic point needs to be made that any reform of public diplomacy organizations and interagency structures must take fully into account the actual and potential role of the Pentagon and the uniformed military in this arena. This includes the sensitive and difficult issues of the role of psychological operations in current doctrine and practice and military–media relations under conditions of protracted low-intensity warfare. More generally, it is essential to keep in mind that any reinvention of public diplomacy cannot involve simply a return to the past but rather must reflect a complete reconceptualization of this subject in the context of a war in which the United States may well find itself engaged for decades to come.

Notes

1 See especially US Advisory Commission on Public Diplomacy, *Consolidation of USIA into the State Department: An Assessment After One Year* (Washington, DC, October 2000); Stephen Johnson and Helle Dale, "How to Reinvigorate US Public Diplomacy," Backgrounder No. 1645 (Washington, DC: Heritage Foundation, April 2003); Council on Foreign Relations Task Force, *Finding America's Voice: A Strategy for Reinvigorating US Public Diplomacy* (New York: Council on Foreign Relations, 2003); Advisory Group on Public Diplomacy for the Arab and Muslim World, *Changing Minds, Winning Peace: A New Strategic Direction for US Public Diplomacy in the Arab and Muslim World* (Washington, DC, October 2003); *Report of the Defense Science Board Task Force on Strategic Communication* (Washington, DC: Department of Defense, September 2004); and Public Diplomacy Council, *A Call for Action on Public Diplomacy* (Washington, DC, January 2005).
2 For a more comprehensive discussion, see Carnes Lord, *Losing Hearts and Minds? Public Diplomacy and Strategic Influence in the Age of Terror* (Greenwood, CT: Praeger, 2006).
3 Clayton D. Laurie, *The Propaganda Warriors: America's Crusade Against Nazi Germany* (Lawrence: University Press of Kansas, 1996).
4 On this history, see especially Walter L. Hixson, *Parting the Curtain: Propaganda, Culture, and the Cold War, 1945–1961* (New York: St. Martin's Press, 1997), chs 1–3; and Wilson P. Dizard Jr, *Inventing Public Diplomacy: The Story of the US Information Agency* (Boulder, CO: Lynne Rienner, 2004), chs 1–4.
5 Quoted in Hixson, *Parting the Curtain*, p. 19.

6 Dizard, *Inventing Public Diplomacy*, ch. 7.
7 Comprehensively listed in Lois W. Roth, "Public Diplomacy and the Past: The Search for an American Style of Propaganda (1952–1977)," *Fletcher Forum* (Summer 1984): 353–96.
8 Panel on International Information, Education, and Cultural Relations, *International Information, Education, and Cultural Relations: Recommendations for the Future* (Washington, DC: Center for Strategic and International Studies, 1975).
9 Quoted in US Advisory Commission on Public Diplomacy, *Consolidation of USIA into the State Department: An Assessment After One Year*, p. 6.
10 See particularly Johnson and Dale, "How to Reinvigorate US Public Diplomacy."
11 See, for example, Peter Carlson, "The USA Account," *Washington Post*, 31 December 2001; and Stephen F. Hayes, "Uncle Sam's Makeover," *Weekly Standard*, 3 June 2002, pp. 22–5.
12 Newt Gingrich, "Transforming the State Department," speech delivered at the American Enterprise Institute, Washington, DC, 22 April 2003, and "Rogue State Department," *Foreign Policy* (July–August 2003): 42–8. For an instructive survey and analysis of Foreign Service attitudes, see Stephanie Smith Kinney, "Developing Diplomats for 2010: If Not Now, When?" *American Diplomacy* (Summer 2000), available at www.unc.edu/depts/diplomat/AD_Issues/amdipl_16/kinney.
13 Author interview with a member of the US Advisory Board for Public Diplomacy, July 2004. Kinney, in "Developing Diplomats for 2010," reports that many Foreign Service officers surveyed would like the Foreign Service to be more like the military in this and other respects.
14 In fact, I supported it conditionally in 1998; see Carnes Lord, "The Past and Future of Public Diplomacy," *Orbis* 42 (Winter 1998): 70–1.
15 Of the major outside studies (see note 1 above), only the Public Diplomacy Council Report calls for the reestablishment of USIA (which it would rename the US Public Diplomacy Agency), although a privatized entity comparable to USIA is recommended by the Council on Foreign Relations Report as well as the Defense Science Board Report. See also Leonard H. Marks, Charles Z. Wick, Bruce Gelb, and Henry Catto, "America Needs a Voice," *Washington Post*, 26 February 2005 (the writers are all former directors of USIA).
16 There may be merit in the idea of establishing an independent, not-for-profit "Corporation for Public Diplomacy" along the lines of the Corporation for Public Broadcasting, as proposed by the Council on Foreign Relations Task Force, but not as a substitute for a robust Educational and Cultural Affairs bureau. In any case, the political carrying costs of such a move could well prove exorbitant.
17 Remarks of Under Secretary of State for Public Diplomacy and Public Affairs Karen Hughes at Town Hall for Public Diplomacy, Washington, DC, 8 September 2005.
18 A planning unit was in fact created within the International Information Programs bureau in September 2004 encompassing the public affairs as well as the public diplomacy function and tasked as the focal point of interagency coordination for the under secretary; a crisis response element is apparently another initiative of Karen Hughes.
19 Little attention seems ever to have been given to this dimension of public diplomacy by the US government. For a useful contrasting perspective, see Mark Leonard, *Public Diplomacy* (London: Foreign Policy Centre, 2002), esp. ch. 3.
20 Reporting responsibilities for the Bureau of International Narcotics and Law Enforcement were only recently transferred from the under secretary for global affairs to the under secretary for political affairs. Other recent developments in this area include the creation of a new deputy assistant secretary of state for democracy within the Bureau of Democracy, Human Rights, and Labor, the creation of an outside advisory committee on promotion of democracy, and a comprehensive internal review of US democracy promotion strategies. See Office of the Spokesman, US Department of State, Fact Sheet, 29 July 2005.

21 A detailed discussion of public diplomacy field operations is beyond the scope of this chapter, but one could imagine an arrangement whereby the public affairs officers and cultural affairs officers in embassies would normally be from USIA, while information officers would be supplied by State.

22 Allan M. Winkler, *The Politics of Propaganda: The Office of War Information, 1942–1945* (New Haven, CT: Yale University Press, 1978).

23 Author interview with senior Senate staff member, July 2004.

6 The one percent solution
Costs and benefits of military deception

Barton Whaley

A shilling's worth of radio countermeasures will mess up a pound's worth of radio equipment.

Dr. Robert Cockburn, Director of the British Telecommunications Research Establishment in World War II

An ounce of deception is worth a 240-pound tackle.

Jake McCandless, Princeton football coach, 1970s

In recent years, deception has come to be recognized almost universally as a force multiplier. Deception almost always yields substantially greater benefits than expected, and the cost of deception is usually far less than we have generally been led to believe. Even at its most costly, deception need not divert more than a very small fraction of the available combat force from the real target. This holds true for the whole range of operations from grand strategy to minor tactics, for conventional and unconventional operations, and for all terrains. Deception has at least an 80 percent chance of yielding surprise, and the payoffs for surprise are impressively high. Surprise multiplies the chances for quick and decisive military success, whether measured in terms of explicitly sought goals, ground taken, or casualty ratios. Deception is seemingly never contraindicated or dysfunctional in the economics of war.

When the costs of being deceived are high, the benefits of detecting deception are correspondingly high. The costs for both deceivers and detectors may be economic – in material or in the time or lives of personnel – or they may be psychological, social, political, or ethical. In this chapter, I explore the costs to the deceiver in material and personnel. Generally, as we shall see, they are quite low.

Assessing the costs and benefits of deception

The economic cost of any deception is best measured by the proportion of the deceiver's resources (money, time, personnel, or apparatus) diverted from the main event. In war, this is the proportion of personnel and material diverted from the usual field of battle. The benefits that may accrue to the deceiver are potentially manifold: quick success at less than the normal cost in personnel and material to oneself and higher cost to an enemy.

In principle, the method of assessment is simple: it is nothing more than old-fashioned operational research, as its British inventors first called it. The problem is getting relevant and more-or-less reliable data to analyze. A deception cost-benefit analysis requires access to good data on both the deceiver and the enemy – the overall combat resources or both sides, the combat resources diverted by both sides as a result of the deception, and the losses to each side as a result of the deception.

Although there is no shortage of case studies that illustrate the general effectiveness of deception (Liddell Hart 1929, 1954; Whaley 1969, 1979), few studies have specifically measured these benefits together with the costs involved. Until we directly cross-check costs and benefits, any conclusions we draw from our case studies, however glowing in their portrayal of the overall success of deception, remain open to question. Moreover, the few existing cost-benefit studies are narrowly focused, concentrating, for example, on deception measures and countermeasures only in aerial combat or during a relative brief time span, such as World War II. This chapter is a preliminary exploration of the many dimensions of the cost-benefits of deception operations.

The general benefits of deception

The theory of deception (presented elsewhere, e.g. Whaley 1969, 1982) predicts several consequences of deception, three of which are covered here: (1) The more intense the deception (i.e. the more of types of ruses used), the greater the surprise – but only up to a point; (2) the greater the surprise, the smaller the force ratio required; (3) the greater the surprise, the fewer the casualties.

More deception yields more surprise – up to a point

The general trend can be seen in Table 6.1, which glosses the minor trends and perturbations in data from World War I through 1968.

Using Whaley's expanded but unpublished DECEPTR database of 232 military engagements from 1914 to 1973, Ronald Sherwin focused on the ninety-three strategic cases (Sherwin and Whaley 1982, 188). Table 6.2 summarizes the results.

Two interesting findings emerge from the table. First, when deception is absent, surprise is rare; conversely, when even a deception employs a moderate number of types of ruses, surprise is common. This finding was expected. The second finding was not: Sherwin found that the "optimal threshold for achieving surprise is between two and three [types of] ruses [out of a possible ten] – probably closer to two [types], depending on which two [types], which target, and in which context." This finding may seem counterintuitive; after all, it might seem reasonable to expect that the use of more deception ruses would bring more surprise. However, one of the more entrancing recent findings of cognitive psychology is that, contrary to popular belief, few decision makers (such as intelligence

Table 6.1 Relationship between intensity of deception and intensity of surprise, 1914–68

Intensity of surprise	Intensity of deception					
	0		1–4		5–8	
	N	%	N	%	N	%
0	53*	72.6	5	6.4	1	6.3
1	6	8.2	9	11.6	1	6.3
2	7	9.6	32	41.0	3	18.7
3	4	5.5	23	29.4	2	12.5
4–5	3	4.1	9	11.6	9	56.2
Total	73	100.0	78	100.0	16	100.0

Note

* This cell represents all 53 type C (non-surprise) examples in the database. While it is not possible to say how representative the figure may be, the bias in selecting type C was to err on the low side.
Intensity of deception refers to the relative level of investment made in terms of material resources, manpower, and even actions in the field (e.g. feints). The most intense surprise occurs when the opponent is unaware that an attack is coming and is deceived even as to the likely places where an attack might arise. The least intense surprise is a situation in which an attack is expected, its location anticipated, but the specific timing is not. Middle levels of surprise generally comprise instances where an attack is expected, but not the location, or vice versa, and instances where the size of the attack may be misestimated.

Table 6.2 Intensity of deception (number of types of ruses used) and whether surprise was achieved in strategic battles, 1914–73

Result	Number of ruses used							Totals
	0	1	2	3	4	5	6	
No surprise achieved	0	0	6	0	0	0	0	6
Surprise achieved	0	3	46	20	12	7	0	87

analysts) wait until all the information is in. Instead, most people reach decisions on the basis of only around three data sets.

Lower force ratios follow from surprise

One of the legacies of the machine-gun- and artillery-dominated Western Front was a "magic" number for success: commanders came to adopt the notion that a 3-to-1 local superiority in the zone of an offensive was necessary for a success-ful attack. Indeed, this rule of thumb was consistent with statistical studies of the *overall* averages of Western Front battles. This bit of Great War doctrine was carried over into World War II and beyond. But how does this 3-to-1 doctrine look in the light of our data on surprise?

First, let's look at the World War I data. As Table 6.3 shows, of the 47 offensive

Table 6.3 Attacker-to-defender force ratio used to gain objectives in World War I

Achievement	Surprise		No surprise	
	N	Force ratio	N	Force ratio
Victory	7	2.1:1	0	–
About as planned	10	1.6:1	5	2.1:1
Below expectations	5	2.8:1	5	1.3:1
Defeat	3	5.9:1	12	1.8:1

Source: Whaley (1969), 199.

Table 6.4 Attacker-to-defender force ratio used to gain objectives, 1919–67

Achievement	Surprise		No surprise	
	N	Force ratio	N	Force ratio
Victory	18	1.2:1	1	2.5:1
About as planned	28	1.1:1	4	1.4:1
Below expectations	17	1.4:1	9	1.4:1
Defeat	4	1.0:1	20	0.9:1

Source: Whaley (1969), 199.

battles for which data were available, the average force ratio of attacker to defender was 2.1 to 1.

There is no obvious pattern here. But before comment, consider the comparable data in Table 6.4 for the half century following the Great War.

Tables 6.3 and 6.4, taken together, support two important conclusions. First, we see strong proof for the notion that a substantial superiority of force is needed, although my data yield force ratios of about 2 to 1 rather than the traditional 3 to 1. However, this general remark holds true only for the more usual type of offensive military operations, that is, those without surprise. Here, there is a fairly direct relationship between force and degree of success: the more the force, the greater the success. Second, and both more important and rather unexpected, is that surprise intervenes to shatter the direct and simple relationship between force and success. In other words, surprise – which is usually a result of deception – is a more important determinant of combat success than brute strength.

Casualty ratios reduced

It follows from deception theory that the more intense the surprise, the more the casualty ratios favor the surpriser. The empirical data available quite emphatically verify this, as Table 6.5 shows.

If deception is not only a main cause but also an enhancer of surprise, then I

Table 6.5 Effect of intensity of surprise on casualties, 1914–67

Intensity of surprise	Mean casualty ratios
0	1:1.1
1	1:1.7
2	1:4.5
3	1:5.4
4	1:4.1
5	1:11.5

Source: Whaley (1969), 194.

Table 6.6 Effects of deception and/or surprise on casualty ratios, 1914–67

	N	Mean casualty ratios
Surprise with deception	59	1:6.3
Surprise without deception	20	1:2.0
No surprise with deception	5	1:1.3
No surprise without deception	40	1:1.1
Total	122	

Source: Whaley (1969), 195.

predicted that the data must show (1) that casualty ratios are substantially greater in cases of surprise with deception than in those of surprise without deception, and (2) that casualty ratios between cases of no surprise and those without deception are similar. Both expectations are verified by the data in Table 6.6.

Cases illustrating costs and benefits

By any purely economic accounting, deception can be relatively cheap – indeed ridiculously cheap. The following case examples illustrate this point.

Case: Normandy, D-Day, 6 June 1944

In absolute economic terms, the most costly deception of all time was Operation Fortitude, the Allied deception operation designed to convince Hitler that the Allied invasion of Nazi-occupied Europe would arrive on the coast of Pas-de-Calais and not further West on the beaches of Normandy. The operation succeeded in keeping 19 of Hitler's divisions tied down in reserve for 66 days after the D-Day landings at Normandy. This benefit to the Allies was enormous, quite possibly making the difference between success and a disaster that would almost certainly have prolonged the war at least a year. But at what cost?

The total cost of all D-Day deception operations was well under 1 percent of that of the invasion force. The breakdown is as follows (I compiled these figures from various sources, beginning in 1972):

- The full-time services of some eighteen deception planners for about a year.
- The part-time attention of perhaps twenty senior staff planners and intelligence officers.
- Operational commitments of roughly 800 radio technicians (some 500 British and 296 Americans) who busily simulated the radio traffic of two phantom armies.
- Several camouflage and construction companies (about 2,000 troops drawn from the Royal Engineers and 379 from the US Corps of Engineers) building dummy installations.
- About 1,000 air force personnel to crew and service the ninety aircraft flying spoof missions on D-Day.
- About 200 sailors for the eighteen small launches involved in misleading the German coastal early warning system.
- Six Special Air Service (SAS) men to parachute on a diversionary mission.
- The ten members of Twenty ("Double-Cross") Committee directing the disinformation campaign of about eleven of the forty German double agents then in Britain.
- Perhaps 100 or so agents and underground members engaged in spreading false rumors on the Continent.
- And, finally, Lt. Clifton James, who was co-opted from the British Army Pay Corps to simulate General Montgomery.

The total diversion of personnel was about 4,500 soldiers, sailors, and airmen, of whom fewer than a quarter were combat personnel. This represents less than 0.2 percent of the available combat personnel, 0.5 percent of the naval craft, and only 0.6 percent of the aerial sorties flown on D-Day.

The total loss among the deception forces was one bomber (its crew safely parachuted), a few tons of aluminum strip air-dropped to deceive German radar, and four of the SAS paratroopers who were taken prisoner and executed (per Hitler's standing order regarding all behind-the-line paratroopers). In sum, the total diversion of well under 1 percent of all combat troops, aircraft, and naval craft at a cost of four personnel casualties, one aircraft, and zero naval craft – in short, the 1 percent solution that inspired the title of this chapter.

Case: RAF bomber offensive: opening the "Window," 1943

The Royal Air Force (RAF) Bomber Command introduced "Window" or "American Chaff" – strips of aluminized paper – into its bag of spoofing tricks on the night of 24/25 July 1943 (this account is drawn from Price 1967, 155, 158; and Webster and Frankland 1961, 156). The target was Hamburg. At 9:55 p.m., the entire Main Force bomber stream of 791 heavy bombers had assembled over the North Sea and set out due east at 225 mph in a typical stream twenty miles wide and 200 miles long. As was typical, 45 bombers aborted for various reasons and returned to base.

German early warning radar spotted the incoming stream at 12:15. At 12:19

the first air raid alarms were sounded and the German night fighters were alerted. Radar tracked the approach toward Hamburg, where the full air raid alarm sounded at 12:31. Simultaneously, each of the Main Force bombers, now twenty miles from the German coast, began dropping Window at the rate of one bundle a minute. To the watching German radar operators – upon whom the circling night fighters were entirely dependent – these 746 bombers suddenly looked like 11,000. The ground controllers were helpless. At 12:50 the bomber stream turned into its final attack bearing, straight to target, twenty-five miles away. At 12:57 the twenty Pathfinder bombers began their target marking and the attack was under way. Only then were the German ground controllers able to order the night fighters to Hamburg. But even then, the night fighters' radars were effectively jammed as they made attack after attack on empty drifting clouds of aluminum foil, while the equally jammed radar of the eighty anti-aircraft guns and twenty-two searchlight batteries sent their flak and candle-power on blind chases. The last of the 740 bombers that actually attacked withdrew at 1:50 after dropping 2,400 tons of bombs, which killed about 1,500 civilians. Only twelve RAF bombers were lost.

Window had been a stunning technological success. On this first use, forty tons of aluminized paper (92,000,000 separate strips) costing £770 subtracted only 1.7 percent of the raiders' bomb tonnage capacity. Yet it very likely saved about 35 British aircraft (with their own 110-ton bomb capacity) worth £1,400,000, plus their 245 aircrew, whose training alone was valued at £2,450,000 (figures calculated from data in Middlebrook 1974; Webster and Frankland [1961] incorrectly calculated a probable saving of seventy-eight bombers).

Case: RAF Window assessment, 1943–4

Window proved highly cost-effective. It was cheap to make and simple to use. The Allies dropped over 10,000 tons of this aluminum foil over Europe alone, simulating 12,500,000 heavy bombers. It cost only about £180,000 to make and subtracted less than 1 percent from the total bomb load of the Allied bombers that flew against Hitler's Fortress Europe under Window protection after mid-1943. Yet it enabled thousands of those bombers and their crews to survive and fight another day. Moreover, for a time, 90 percent of Germany's 7,800 invaluable high-frequency experts were diverted to the development of an effective electronic countermeasure.

The hypothetical savings achieved by Window cannot be calculated exactly, because the ever-changing factors (German countermeasures, new Allied tactics, other Allied deception methods, etc.) make overall before-and-after studies incomparable. Even rough estimates would require an exceedingly elaborate statistical operations research study. However, the available fragmentary evidence strongly suggests that the Germans were never able to overcome the effects of Window, as shown in Table 6.7.

Perhaps as good a summary statement on Window as any is Churchill's: "Its effects surpassed expectations.... For some months [after Hamburg] our bomber

Table 6.7 Percentages of bomber losses before and after Window raids began, 1943–4

Target city	Dates	Percent lost with Window	Percent lost pre-Window	Percent saved
Hamburg	24 July–3 August 1943	2.8	6.1	3.3
Essen	25/26 July 1943	3.4	5.4	2.0
Berlin	23 August–3 September 1943	7.2	c.10.0	c.2.8
Berlin	18 November–24 March 1944	5.4	c.10.0	c.4.6

Sources: Compiled and calculated from Webster and Frankland (1961) and Price (1967).

Table 6.8 RAF bomber command order-of-battle, 30 March 1944

Mission	Squadrons	
	N	%
Main force bombers	67	85.5
Secret ops	2	3.5
Ferrets	2	3.5
Diversionary	4	5.5
Jammers	1	1.5
Total	76	99.5

Source: Compiled from Middlebrook (1974), 322–5.

losses dropped to nearly half. And up to the end of the war, although the German fighter planes increased fourfold, our bomber losses never reached the same level that had been suffered before '*window*' was used" (Churchill 1950, 288–9).

Case: Assessing RAF Bomber Command deception cost, 1944

Another measure of RAF Bomber Command's deception costs is to calculate the percentages of its order-of-battle devoted to deception operations. As Table 6.8 shows, the total percentage cost in squadrons devoted to deception – diversionary plus jammers – was 7 percent of the total number of squadrons. However, as each of these deception squadrons cost only about half that of a Main Force squadron in aircraft and number of aircrew, the adjusted cost of the deception squadrons is only around 3.5 percent of Bomber Command's total strength.

Case: Maj. Ingersoll and the Battle of the Bulge, 1944

Urgent improvisation of a deception plan in its extreme form was suddenly needed during the critical Battle of the Bulge in December 1944. The German army had just broken through the Allies' weak point in a surprise attack. One measure of this surprise was that Col. William "Billy" Harris, who headed Gen.

Omar Bradley's 12th US Army Group's Special (i.e. deception) Plans Section, was away on other matters. Thus, at this crucial moment, command devolved upon his deputy, forty-two-year-old Maj. Ralph Ingersoll, a peacetime newspaper publisher. Ingersoll would be forced to improvise. His plan would prove to be a masterpiece of cost-beneficial deception. I call it the "Two Pattons Ruse." (This account is drawn from author's interview with Ralph Ingersoll, 12 May 1973; and Hoopes 1985, 304).

On 19 December, Gen. Eisenhower went to Bradley's rear headquarters at Verdun to meet Bradley and Patton and decide on a strategy to stop the German offensive. Ingersoll was present at the meeting. The battle-eager Lt. Gen. Patton assured Eisenhower and Bradley that he could rush his powerful Third Army north and stop the enemy advance. The two senior commanders were doubtful because of the immense logistical problems, but desperate measures were needed, and they unleashed Patton.

Urgency demanded that Patton's movement orders to his corps and division commanders would have to go out over his radio nets "in clear" (not coded), despite the certainty that the German signals intelligence teams would overhear them.

Everyone involved knew it was too late to mount any conventional deception operation to cover this counterattack. Nevertheless, Gen. Bradley turned to Maj. Ingersoll and asked, "Is there anything you and your people can think of to throw the Germans off – about where they [Patton's divisions] are going to strike?" Ingersoll replied, "Yes, sir. We can do …" – he hesitated – "something." Bradley was momentarily silent before simply ordering, "Then do it" (Hoopes 1985, 304).

Ingersoll later said that the idea had come to him in a flash. Realizing during the meeting that time would be too short for any conventional deception, he'd been trying to think of some alternative. His mind lit on the recent example of the vast confusion to German military intelligence created on D-Day by the unintentional scattering of the two American and British airborne divisions. With this thought in mind, Ingersoll suddenly conceived – indeed invented – the unprecedented ruse of the Two Pattons.[1] He code-named the operation Kodak because he intended to confuse the enemy by giving them a "double-exposure" (*Official History of the 23rd Headquarters Special Troops*, 1945, 23).

The only resource available to Ingersoll was the Special Plans operational deception unit, the 23rd Headquarters Special Troops. It was too late to deploy their usual panoply of visual and aural spoofs. Radio deception alone would have to do the job. Accordingly, nineteen officers and twenty enlisted men were detached from the 23rd's Signal Company, rushed forward, and dispersed among advance units of Patton's Third Army. Another eleven officers and 205 EM provided support farther back. It was, as the official history called it, "a big (29 sets), loose radio show" (*Official History 23rd*, 1945, 23; also Enclosures I and V).

These radio operators normally spent their time simulating "notional" (fictitious) US Army units in a largely successful effort to dupe the enemy into wasting precious resources by either advancing or retreating from these phantoms. But now, for two days, 22 and 23 December, Ingersoll had them imitate two real units – the 80th Infantry Division and the Fourth Armored Division, which were spearheading Patton's advance. The 23rd's bogus radio show was timed to start when these division radio nets began broadcasting in clear.

The real 80th was getting ready to jump off due north from Luxembourg, and the real Fourth was preparing to roll up from Arlon to effect its historic relief of the 101st Airborne Division, which was trapped in Bastogne. "The 23rd mission was to show ... the presence of the 80th Infantry and 4th Armored Divisions slightly northeast of Luxembourg and in position to forestall any German plans of extending their counterattack southwest through Echternach" (*Official History 23rd*, 1945, 52).

In effect they had presented German army intelligence with two Gen. Pattons approaching the Bulge from slightly different directions. Of course, the Germans quickly understood the radio game being played against them, but they were still dazzled by the two fake divisions advancing eastward of the two real ones. In the event, the Germans were able to keep track of only one of Patton's advancing divisions. All others were either lost entirely on the German battle maps or, worse, mislocated. Consequently Patton was able to gain a substantial tactical surprise, which helped him break the back of the Wehrmacht's last offensive.

Long after the war, when I interviewed Ingersoll about his role in military deception, the Two Pattons story – unpublished at that time – emerged only as an afterthought. Ingersoll mentioned it, saying apologetically that, while proud of his improvisation, he didn't classify it as "deception," which he defined as making the victim certain but wrong. Instead he chose to make the enemy merely uncertain and confused and to hope for the best. Ingersoll's unique ruse does indeed fit one category of deception, the one called "dazzle" in the Bell-Whaley typology (Whaley 1982, 184; dazzle is the same as the "A type" [ambiguity-increasing] deception in Daniel and Herbig 1982, 5).

In sum, the benefits of Operation Kodak were a major defeat for the enemy's army and saving the Allied drive into Germany a delay of at least a couple of months. The cost was the employment of thirty officers, 225 enlisted men, and twenty-nine radio sets – under 0.1 percent of Bradley's 12th US Army Group's personnel and none of its combat personnel.

Case: Overall cost-benefit assessment of the US Army's First Tactical Deception Field Unit, 1944–5

This is a rare case where we have enough data on a specific military deception operations unit to assess its overall benefits and costs with some precision.

In January 1944, the 23rd Headquarters Special Troops was activated at Camp Forrest, Tennessee. This was the first self-contained US military organization specially and solely designed for tactical deception (this account is drawn from *Official History 23rd*, 1945; Mendelsohn 1989; Kneece 2001; Gawne 2002; Gerard 2002). The 23rd was a reinforced battalion-sized unit, commanded throughout its twenty-one-month existence by Col. Harry L. Reeder, who had been a regular officer since before World War I and had considerable tank experience and an unconcealed contempt for this noncombat assignment. At full strength, the 23rd mustered 1,106 troops (eighty officers, three warrant officers, and 1,023 enlisted men). It was structured as follows: Headquarters Company had 108 officers and men who exercised overall direction. The 603 Engineer Camouflage Battalion Special (with 379 officers and men, mainly New York and Philadelphia artists with an average IQ of 119), handled conventional camouflage and the inflatable rubber dummy tanks, trucks, and guns. The Signal Company Special (296 officers and men, of whom 196 were radio operators) handled the radio deceptions. The 3132 Signal Service Company Special (147 officers and men), the first such unit in the US Army, handled sonic deception. The 406 Engineer Combat Company Special (168 officers and men) handled security and rough jobs in general.

What was the 23rd's learning curve and what was its final score? Consider first the learning process, as outlined in Table 6.9. Here we see a pattern of ups

Table 6.9 Results of plans and operations of the 23rd Headquarters Special Troops in France and Germany, 1944–5

Date	Operation or plan	Result
1944		
6 June	No name	Aborted (situation found unsuitable)
7 June	Troutfly	Aborted (change of battle plan)
1–4 July	Elephant	Partial success
9–12 August	Brittany	Success
20–27 August	Brest	Partial success, partial backfire
1 September	No name	Aborted (too late)
15–22 September	Bettembourg	Partial success
4–10 October	Wiltz	Unknown
10 October	Vaseline	Aborted (change of battle plan)
2–10 November	Dallas	Probable success
3–12 November	Elsenborn	Partial success
4–9 November	Casanova	Success
6–14 December	Koblenz	Partial success
21 December	Koblenz II	Aborted (liaison officer lost trailer)
22–23 December	Kodak	Success exceeding expectations
28–31 December	Metz I	Unknown
1945		
6–9 January	Metz II	Unknown
10–13 January	L'Eglise	Unknown
17–18 January	Flaxweiler	Probable success
27–29 January	Steinsel	Probable success
28 January–2 February	Landonviller	Unknown
1–4 February	Whipsaw	Partial success
13–14 February	Merzig	Failed
1–11 March	Lochinvar	Partial success
11–13 March	Bouzonville	Partial success
18–24 March	Viersen	Success

Table 6.10 Overall results of the 23rd Headquarters Special Troops' deception plans, 1944–5

Result	N	%
Success	4	15.5
Partial or probable success	11	42.5
Failure	1	4.0
Unknown	5	19.0
Aborts	5	19.0
Total	26	100.0

Sources: Generated from *Official History 23rd*, 1945, Enclosure I; modified by details in pp. 8–31; and corrected by upgrading two cases from more recently available data, one from "unknown" to "successful."

and downs but a gradual overall improvement in performance. Aborted operations became less common because of closer liaison and coordination with the combat units being supported. Simulation of the army divisions being portrayed grew more realistic, as measured by the number and quality of their various signatures.

What was the final score? As Table 6.10 shows, the 23rd generated a total of twenty-six tactical deception plans in support of Gen. Bradley's twenty-nine US divisions during its ten operational months from Normandy in June 1944 until March 1945. Of these twenty-six plans, five were aborted and twenty-one became actual operations. Of the latter, four were entirely successful, eleven were at least partly successful, and only one failed completely. Of the five operations whose results are unknown, the very lack of evidence suggests failure or only negligible success. On that pessimistic assumption, we get fifteen "successes" against six "failures" plus five aborted operations – a cost-effective record given the small size of the unit and the nature of the German reaction to its deceptions.

We can't accurately calculate how many lives the 23rd saved during its eleven months in action. From the D-Day landings on 6 June 1944 until the German surrender on 8 May 1945, 135,000 Americans were killed and 450,000 wounded. One historian of the 23rd concluded, "It is not much of a stretch to say that its other twenty operations could have brought the total [saved] to about 40,000 Allied lives. Each time German divisions were frozen in place, it meant that they were not killing Americans elsewhere. It also meant they themselves suffered fewer casualties.... The Ghost Army was just a small unit of 1,100 men among 5,412,219 Allied troops under Eisenhower during the war in Europe. But they projected the work of thousands" (Kneece 2001, 280).

Defensive deception

Deception is cost-effective for the initiators and attackers. Less well appreciated is that it can also work well for the defender. Two spectacular examples are the 1915 evacuation of Suvla Bay and Anzac Beach and the 1916 evacuation of Cape Helles.

Case: Evacuation of Suvla Bay and Anzac Beach, 20 December 1915

This was a case where even the enemy could not help but admire the deception: "So long as wars exist, the British evacuation of the An Burnu [Anzac] and Anafarta [Suvla] fronts will stand before the eyes of all strategists of retreat as a hitherto unattained masterpiece" (*Vossische Zeitung*, 21 January 1916).

On 7 December 1915, the British Cabinet agreed to liquidate all of the Gallipoli enterprise except the tip position at Cape Helles (this account is drawn from Whaley 1969, A56–A62). This difficult and embarrassing decision had been made with the concurrence of the War Committee, Lord Kitchener (the minister for war), Lt. Gen. Sir Charles Monro (Ian Hamilton's successor as commander in the Aegean), all of Hamilton's former staff (now under Monro), and two of the three corps commanders on Gallipoli. The now isolated group favoring further escalation at Gallipoli was largely limited to Churchill, Commodore Keyes, Lord Curzon, the replaced Hamilton, one able Gallipoli corps commander (Birdwood), and the secretary of the War Committee (Colonel Hankey).

Moreover, the decision to evacuate was made in full concurrence with predictions of appalling numbers of casualties during the withdrawal phase. As early as 12 October, Hamilton had informed Kitchener (and through him, the War Committee) that an evacuation would entail 50 percent casualties and loss of all guns and stores. Although this was an exaggeration designed to forestall a decision to cancel Hamilton's pet operation, even he expected 35 to 45 percent casualties in such an event.

In late October, Hamilton's former staff, although generally biased in favor of withdrawal, presented the newly arrived Monro with a careful estimate that reckoned on a 50 percent loss in troops and 66 percent in guns. Monro notified Kitchener on 2 November of his own estimate of losses of 30 to 40 percent of both men (i.e. some 40,000) and material. On 15 November, Kitchener more optimistically reported to London that he expected about 25,000 casualties (i.e. 25 percent), which was also then the reduced estimate of the local staff. On 22 November, the British Army General Staff reported to the War Committee its estimate of 50,000 casualties (i.e. 50 percent).

These dire predictions were naturally based on the assumption that the evacuation would be contested. Lord Curzon, the Lord Privy Seal, in a memorandum circulated to the Cabinet on 25 November, depicted the operation in terms of doom:

> [The] evacuation and the final scenes will be enacted at night. Our guns will continue firing until the last moment ... but the trenches will have been taken one by one, and a moment must come when a final *sauve qui peut* takes place, and when a disorganized crowd will press in despairing tumult on to the shore and into the boats. Shells will be falling and bullets ploughing their way into this mass of retreating humanity.... Conceive the crowding into the boats of thousands of half-crazy men, the swamping of craft, the nocturnal panic, the agony of the wounded, the hetacombs of slain.
>
> (Quoted in Ronaldshay, 1928, 130–1)

It is interesting that the differences among these various casualty predictions were directly associated with each predictor's bias for or against evacuation. Advocates of escalation pressed forward the more pessimistic figures, and advocates of evacuation advanced the optimistic estimates. Moreover, those individuals who changed their position on evacuation simultaneously adopted the psychologically appropriate statistics. After evacuation had been ordered by London, most of the local staff officers suddenly revised their private estimates downward to a comforting 15 percent. Thus, casualty estimates, far from being rational military calculations, were used by the military and political professionals alike as a political tool and a psychological crutch.

In planning the deception for the evacuation, Monro rejected the conventional feints or demonstrations as a supplement to the various ruses at the beachheads themselves; a feint could well arouse the Turks' suspicion, with disastrous results. Instead, Monro "came to the conclusion that our chances of success were infinitely more probable if we made no departure of any kind from the normal life which we were following both on sea and on land" (Monro despatch of 6 March 1916, as quoted in Callwell 1919, 276–7).

Portents of the decision to evacuate Gallipoli had been gradually accumulating before the public and enemy eye alike. Thus, the commander in chief of the expeditionary force, Hamilton, was relieved on 14 October. That same day, withdrawal was first openly advocated in the House of Lords by Milner. On 2 November, in the House of Commons, Prime Minister Asquith bluntly admitted the failure of the recent offensive. The announcement on 11 November of a reconstituted War Committee (to replace the Dardanelles Committee) revealed publicly that the two leading public proponents of the adventure, Churchill and Curzon, had been dumped.

Finally, on 18 November, Lord Ribblesdale, an outspoken advocate of withdrawal, asserted in the House of Lords that General Monro had "reported in favor of withdrawal from the Dardanelles, and adversely to the continuance of winter operations there." It is interesting that on learning of Milner's statement, the Turks discredited it as a deliberate effort at deception. They fully expected the Allies to remain on Gallipoli.

It was decided that the evacuation would take place at night, although some troops were embarked under cover of tarpaulin in daylight. Aviation patrols were depended on to keep the few German reconnaissance aircraft from getting too close a look at the denuded positions, particularly on the last day, when no fewer than five Allied aircraft maintained almost continuous air cover.

The monumentality of this task and its success are summarized in the timeline of the evacuation presented in Table 6.11. Yet it was completed without raising the suspicions of the Turks or the German commander, General (and Turkish field marshal) Liman von Sanders.

The evacuation on the last night was effected with no dead and only five wounded and with only ten out of 200 guns abandoned. Of some 5,000 draft animals, only fifty-six mules were left. Even most of the 2,000 carts were salvaged.

A measure of the effect of surprise can be dramatized by comparing predicted

Table 6.11 Diminishing strengths at Suvla and Anzac, 8–20 December 1915

Date and time	Suvla		Anzac		Total	
	Men	Guns	Men	Guns	Men	Guns
December						
8	43,000	90	40,000	110	83,000	200
10	?	?	?	?	80,000	?
17	?	?	?	?	44,000?	?
18/0600	16,000	?	22,000	?	38,000	?
2000	16,000	?	22,000	?	38,000	?
19/0530	9,000	16	11,000	10	20,000	26
2000	9,000	16	11,000	10	20,000	26
2200	2,000?	0	4,000?	10	6,000?	10
20/0100	1,322+	0	2,000?	10	3,500?	10
0400	550	0	800?	10	1,350?	10
0415	200	0	0	10	200	10
0515	0	0	0	10	0	10

Table 6.12 Predicted and actual losses in the Suvla and Anzac evacuation

		Predicted losses					
		Pessimistic		Optimistic		Actual losses	
Men and materials	Total force	N	%	N	%	N	%
Men	83,000	40,000	50.0	12,500	15.0	5	0.01
Guns	200	100	50.0	30	15.0	10	5.0
Horses and mules	5,000	2,500	50.0	750	15.0	56	1.1

and actual losses (Table 6.12). Even the "optimistic" predictions exceeded actual losses by a factor of 2,500 for troops, a factor of 3 for guns, and a factor of 14 for draft animals.

The cost-effectiveness of surprise was so little understood by many that one common but false rumor credited the success of this bloodless coup to the Turks' having been bribed to permit the evacuation without opposition.

Case: Evacuation of Cape Helles, January 1916

The success of the evacuation of the beachheads at Suvla Bay and Anzac Beach hurried the final decisions for the complete liquidation of the Gallipoli enterprise (this account is drawn from Whaley 1969, A63–A65). The only remaining beachhead was at Cape Helles, at the very tip of the peninsula. On 23 December, the British War Committee recommended withdrawal; the Cabinet agreed to it on the 27th, and General Monro, the commander in chief at Gallipoli, was notified the next day.

The situation on Gallipoli again seemed desperate. The remaining Allied force of four divisions (35,268 troops) were crowded into the twelve square mile beachhead at Cape Helles. Facing them was a vastly larger force of Turks directed by General von Sanders. With the British evacuation of Suvla and Anzac completed, von Sanders was concentrating his entire force of twenty-one divisions (120,000 troops) against the remaining Allied position. It was certain that a contested evacuation would be costly, and perhaps even disastrous.

Could a second surprise withdrawal be made? Again General Monro chose to depend on simulating normalcy and avoiding any feints (despatch of 16 March 1916, quoted in Callwell 1919, 310). The basic plan that had worked so well a fortnight before was used again, although this time with one new technique: the evacuation of Cape Helles was carried out under the cover of reinforcement and relief. This deception succeeded so well that even the troops on the beachhead did not fully recognize what was under way until some time later. As some relief and reinforcement had been in effect since early December anyway, "normalcy" at Helles already meant much coming-and-going. Now, however, beginning 29 December, the pattern of relief shifted slightly to bring in only units of IX Corps, which had experience in the special techniques of evacuation, having just extricated itself from Suvla. Published orders were given out that VIII Corps was being relieved by IX corps. Tables 6.13 and 6.14 summarize the evacuation and the casualties in the Cape Helles evacuation.

Table 6.13 Diminishing strength at Cape Helles, 29 December 1915–9 January 1916

Date and time	Troops	Guns
29 December	40,000	150
31 December	35,268	140
2 January	31,000	?
7 January	19,418	63
8 January 0500	17,118	54
2000	17,118	54
2330	9,918	17
9 January 0145	3,877	17
0300	200	17
0345	0	17

Table 6.14 Casualties at Cape Helles

Date	British		Turks
	Men	Guns	
7 January	164	0	"Heavy"
8 January	Few	0	Few
9 January	5	17	None

Camouflage and decoys

The hiding of certain targets and the deliberate show of others to work as decoys
are common components of deception.

Case: Col. Turner's department: British airfield and port camouflage and decoys, 1939–45

During World War II the British built a total of 797 separate decoy sites on their
island plus a few on the Normandy beachhead and in northern France. All of
these sites were designed and installed by what, for security purposes, was
called simply and officially "Colonel Turner's Department." Among these sites
were 231 somewhat elaborate dummy airfields; 237 civil "Starfish" fire sites to
simulate eighty-one cities; numerous decoys near twenty-nine ports; army
decoys near forty-four posts; and eleven simple dummy buildings (Dobinson
2000, 241–89, and throughout his main text). Unfortunately, I've been unable to
find specific statistics on the personnel and material costs of this British effort.
However, the famously frugal-minded Col. Turner made extensive use of
second-line military personnel, civilian volunteers, and co-opted motion-picture
designers and carpenters from London's suburban Shepperton movie studios.

The British camouflage effort on the post-D-Day Normandy beaches was
highly effective. Four sections of Gold Beach, where the British XXX Corps
landed, were covered by dummy targets erected by a company of Royal Engineers that was supplemented by a small detachment of RAF camouflage and
decoy personnel from the now redundant Starfish night decoys in Britain. These
Normandy decoy sites, which operated from D-Night (6 June) until 15 September, were thought by local reporting officers to have diverted more than half the
German bombs, although, admittedly, German bombing had been lighter than
expected (Dobinson 2000, 202–3).

The overall British camouflage effort is summarized in Table 6.15. Given
Col. Turner's meticulous accounting methods, these statistics are very likely
underestimates (Dobinson 2000, 209–13, 290–8).

Table 6.15 Calculation of British decoy effectiveness in diverting German bombs,
1939–45

Target	Decoy type	Number of attacks	Average tons diverted	Total tons diverted
Airfields	Day decoys ("K sites")	47	2.5	118
Airfields	Night decoys ("Q sites")	521	1.7	860
Various	Night "QF" fires and "QL" lights)	173	1.5	265
Buildings	Day	9	1.3	12
Cities	Night decoys (Starfish)	119	5.6	968
Total		869		2,223

Source: Based on Dobinson 2000, appendix III, 298.

Table 6.16 Estimates of German bomb tonnage diverted by British decoys, 1939–45

Total German bomb tonnage	Total tons diverted by decoys	% tons diverted by decoys
56,000 by night	2,092	3.7
12,500 by day	129	1.0

Distilling these data, we obtain the figures in Table 6.16, which are conservative overall estimates of German bomb tonnage diverted for the entire war. The best adjusted overall proportion of tonnage diverted by decoys is probably 5 percent, and for crucial years and targets the proportions occasionally rise as high as 10 percent. If we accept the reasonable 5 percent estimate, this converts to a saving by moderately inexpensive decoys of more than 3,000 injured and 2,500 dead (Dobinson 2000, 210–13).

The bottom line is that for a modest cost in personnel and material, the British camoufleurs achieved a small but evidently cost-effective saving in casualties and damage to airfields, factories, cities, and port facilities.

Lures

Unlike a decoy, which draws attention away from a nearby real target, the lure stands alone. A lure is a worthless installation designed to draw attack by simulating high value or to provoke the opponent to rise to an unequal fight. Unlike the decoy, the lure is vigorously defended but on a killing ground of the defender's choice. Its purpose is to create a battle of attrition where the casualty ratios greatly favor the defender.

However, unless the lure is simultaneously an ambush, it is in effect no more than a form of attritional warfare and thus is profligate in material and personnel. In absolute losses, the lurer should expect to suffer about as much damage as he inflicts. This tactic of attrition is commonly justified on the grounds that it yields a relative advantage in mutual slaughter. This is a weak argument, for four reasons. First, advocates of attritional warfare often simply neglect taking into account the ability of their opponent to replace his losses; that is, time – and attrition is a matter of time – often works to the advantage of the opponent. Second, the time factor also gives the enemy a good opportunity to see through the intent and adjust his tactics accordingly. In other words, attritional warfare, being intrinsically slow paced, is seldom able to incorporate the most cost-effective factor – surprise. Third, the same result to the enemy at less cost to oneself can usually be achieved by more imaginative deception operations. And fourth, in practice, the attritional lure often backfires.

Lures apply across the range from minor tactics to grand strategy and from asymmetrical to conventional combat. The following is a successful operational-level case.

Case: The German ruse against Operation Crossbow, 1944

Operation Crossbow was Britain's effort in World War II to counter Hitler's super "Victory" weapons (this account is drawn from Irving 1964, 307–8). British intelligence, particularly its photo interpreters (PIs), took pride in their ability to locate the various launch sites before activation. This target information was used to plan 69,000 Allied aerial sorties that unloaded 122,000 tons of bombs at the high cost of 450 aircraft and 2,900 airmen. Yet, more than a quarter, and probably more than half, of this prodigious effort was wasted on obsolete targets. Indeed, not only was the entire 28,000-ton effort from January through May 1944 wasted, but at no time were delays imposed on German launch site preparations before they actually opened fire. This sorry record for the PIs was not only unsuspected at the time, it is still largely overlooked by historians and memoir writers.

The Germans' original "Flying Bomb" installations (which the British PIs called "ski sites," "large sites," and "supply sites") had become obsolete that January. They were henceforth superseded by the well-camouflaged "modified sites," which the PIs failed to detect until the sites began firing in June. Instead of simply abandoning the old sites, the Germans continued to simulate normal activity around them. For five months, the Allies expended their entire Crossbow attack on these now worthless targets, one of the larger of which had been appropriately renamed Concrete Lump. Even after the new "modified sites" revealed themselves in June through their activity, Crossbow still gave the old sites high target priority.

The PIs were adept at identifying *decoys*, that is, dummy installations simulating real ones at alternate locations, but they were quite unable to identify *lures*. A decoy simulates a real target but in an alternate place, and exaggerates value by simulating high payoff or value. A decoy is a pure case of dazzle in that it creates ambiguity of place. A fisherman's lure does not simulate another real object; it merely draws attention to itself in order to draw the fish to it. Even the hunter's "decoy duck" is actually a lure.

The RAF PIs were not only unable to identify lures, they were seemingly unaware of the possibility of lures. Lures are much more difficult to detect than decoys because they are real (not dummy) objects; the lure is the real thing – only its function is faked. Thus camera imagery alone is not enough to uncover the hoax. Here, human intelligence trumps imagery intelligence.

Lures that backfired

As noted earlier, the attritional lure often backfires, as was the case for the Germans at Verdun and the French at Dien Bien Phu. In both cases, the strategy was effective at the outset, but when the tide turned and the lurers' losses mounted, they were unable to relinquish the prized real estate they held, and in the end suffered enormous losses.

At Verdun, the strategy failed because General Erich von Falkenhayn had kept his plan secret from senior headquarters, from parallel headquarters, and from the Austrian allies – even his own artillery adviser was not informed until it was too late to appropriately modify the artillery program (von Falkenhayn 1920, 255, 264–5). Because his superiors and colleagues had no idea that his intention was a strategy of attrition, they soon joined the French in viewing Verdun as a symbol, a precious trophy that must be held at any cost; as a result, that blood-soaked patch of land become a killing ground for the Germans as well. When the battle ended ten months later, the total casualty rates only slightly favored the Germans – 337,000 Germans to 377,000 Frenchmen, or 1 to 1.1, compared with a ratio of 1 to 1.5 in other sectors of the Western Front during the same period.

General Henri Navarre very deliberately committed major elements of his French Army in Indochina to Dien Bien Phu. The French had taken the isolated mountaintop prize of Dien Bien Phu from the Viet Minh by airdrop in November 1953 and gradually increased its garrison to a peak strength of 13,000. Navarre's plan was to lure Giap into adopting a frontal attack or siege tactic; the bait had to appear temptingly weak while in fact being strong – a standard tactic among the French (see Fall 1967, 5, 49). Giap accepted the challenge and mounted a siege, bringing in more and more troops, just as Navarre had wanted. With complete command of the air, Navarre assumed he could reinforce and resupply as long as necessary. The early counterattacks were easily repulsed, with high casualty ratios (well over 3 to 1) for the Viet Minh.

All was going according to Navarre's plan. But then, as with Verdun, Dien Bien Phu started to become a household word not just in France but throughout the world; this trivial mountaintop was becoming transformed into a symbol of prestige and honor for France – and for Navarre. Meanwhile, everything was also going according to Giap's plan. Instead of trickling in one unit after another to be destroyed piecemeal, he gradually built up the besieging force to 50,000 combat troops and 55,000 support troops, winning the race of men and material. As the defense perimeter was squeezed tighter and tighter, the French helipads became vulnerable to artillery fire. Air resupply had become limited to airdrops, and reinforcement as well as evacuation had become impossible. The French were now caught in their own trap, and Dien Bien Phu was overrun six months after they took it, with losses nearly equal on both sides (18,916 French to 22,900 Viet Minh).

Diversionary attacks

The diversionary attack is one type of military deception operation that often proves quite costly, because it sends real combat units into real battle. At its worst, it is the most primitive form of deception. The diversionary attack exemplifies the relatively low level of deceptive thinking among most British, French, and German generals in World War I and the unprepared Russians and Americans early in World War II. For example, in 1917, in the three-day Battle of Scarpe and Vimy Ridge, twenty-two British and Canadian divisions were sent over the top, costing them 18,000 casualties and inflicting a slightly lower number of casualties on the German defenders. But this was merely a *limited diversionary attack* to cover the real Franco-British forty-six-division attack that followed elsewhere (Whaley 1969, A85–A87).

Although this strategy is common during war, such profligacy is not simply inelegant, it is seldom, if ever, necessary. Any number of other deceptions will very likely cost substantially less while providing a substantially higher payoff.

Comparative costs of operations with little or no deception

Most of the military cost-benefit statistics available can at best only compare data from before and after deception was instituted and hope that the class of all exceptional events were typical. A true scientific experiment would examine a stream of events and omit all deceptive measures in one or more randomly selected events. Of course, this normally is not an option. However, by pure chance, just such an "experiment" occurred once during the British bomber offensive against Germany during World War II. This instance was caused by an incredible break in standard operating procedures that produced the disastrous raid on Nuremberg on the night of 30/31 March 1944.

Case: RAF bomber raid on Nuremberg, 30/31 March 1944

That raid was typical of Main Force operations during the so-called Battle of Berlin, which comprised 35 raids on cities throughout Germany during the four and a half months from mid-November 1943 through March 1944 (this account is drawn from Middlebrook 1974; Webster and Frankland 1961, 192–3, 207–9; Price 1967, 196–8; and *The Royal Air Force* 1954, 28; see also Bekker 1968, 339–40; Irving 1964, 54–55; and Verrier 1968, 293). It was typical in all major particulars except one – deception. Strangely, Bomber Command omitted virtually all its best ruses. The Main Force of 795 heavy bombers flew an almost direct course toward their target, with only two minor turning points. None of the normal diversionary operations were undertaken. (Later, Bomber Command limply attributed this to "conditions over the North Sea" without explaining why other routine operations were able to proceed there.) A mine-laying raid by fifty Halifax bombers in the Heligoland Bight did not confuse the German ground controllers. Nor did the small-scale spoof raids on Kassel, Cologne, Frankfurt, and other minor targets, because the Germans immediately recognized the raiders as unthreatening Mosquito bombers.

An interceptor force of 246 night fighters was scrambled and waiting when the Main Force crossed into German airspace at midnight. In the clear night sky, lit by a half moon, the interceptors could see the bomber stream, so they no longer needed their ground controllers' instructions, which in any case were now being jammed. During the hour and a half the bomber stream was fighting its way the 250 miles into Germany, the night fighters shot down about fifty-eight bombers. Given the furious distraction of the aerial combat and the high tail winds, about twenty-seven other bombers aborted or lost their target, and the rest were so scattered that the ground controllers did not actually identify Nuremberg itself as the target until two minutes before the attack began by the 710 remaining bombers. This element of fortuitous surprise did catch seven squadrons of the single-engine "Wild Boar" interceptors hovering outside Berlin and Leipzig. However, twenty-one squadrons (75 percent of all scrambled interceptors) were clinging to the bomber stream and destroyed another twenty-two bombers over Nuremberg and a final fifteen on the homeward flight. In all, the British raiders suffered ninety-four bombers lost, an appalling rate of 11.8 percent. In addition, twelve bombers were damaged beyond repair and fifty-nine were moderately damaged. Nuremberg received only minor bomb damage.

If we merely take the "official history" statistics and recast them as in Table 6.17, the key point is made. During this same period, Mosquito light bombers flew 2,034 diversionary sorties, losing only 10 aircraft, or 0.4 percent (Webster and Frankland 1961, 199).

Three points are clear. First, the confusion and surprise produced by effectively applied deceptions yield average loss rates of only about 5 percent (and those other 34 raids included some where, despite good deception, the German ground controllers or some interceptors accidentally guessed the real target). Second, where deception was poor and surprise was slight, the loss rate soared to nearly 12 percent. Third, the costs of the electronic and spoofing operations themselves were small relative to their effect on the Main Force efficiency and loss rates.

This period (November–March) was one in which the 5.2 percent attrition rate in heavy bombers was "acceptable" to Bomber Command because replacement of planes and crews could keep up – indeed, the total daily average number of available heavies (with trained crews) increased from 864 to 974 over those five months. However, an attrition rate averaging 5.7 percent or more in that same period (i.e. costing an additional 110 planes) would have been "unacceptable." In other words, the already narrow margin of operational existence of Bomber Command in that period must be entirely credited to the deception planning, which was then "saving" about 7 percent (give or take a half percent) of the sorties, or at least 1,300 heavy bombers – more than the total force.[2]

Table 6.17 British heavy bomber losses in the Battle of Berlin, 18 November 1943–31 March 1944

Target	Number of raids	Number of sorties	Bombers lost	
			N	%
Nuremberg	1	795	84	11.8
Berlin	16	9,111	492	5.4
All other targets in Germany	18	10,318	461	4.5
Total	35	20,224	1,047	5.2

It is commonly argued that the strong needn't bother with deception. That may be so, but only in two circumstances – both illustrated by this case: first, when, in the short run, the stronger party isn't bothered by its losses; and second, in the long run, when it isn't concerned that the attrition rates from successive battles will exceed the rate of replacements.

Small-unit asymmetric combat: low-tech, high-deception units versus high-tech low-deception units

The accidental experiment described in the case example above and, more generally, in studies by Liddell Hart (1929, 1941, 1954), Jones (1957), Davies and Harris (1988), and Whaley (1969, 1973, 1979) have demonstrated the high benefit-to-cost ratio of military deception at the levels of grand strategy, strategy, and grand tactics (operational art) that cover entire wars and major battles. These studies suggest that the use of deception confers benefits across the entire strategic-to-tactical continuum. I've long suspected that the many studies that set up a rigid dichotomy between "strategic deception" and "tactical deception" perpetuate a largely false notion. But is this hypothesis on the right track? Since 1969 I've seen many small-unit tactical cases that support this conclusion, and I've found no counterexamples. But this is only anecdotal evidence. We need a robust, systematic operations research–type study focused on this question, on the model of Richardson (1960), Whaley (1969), Sherwin and Whaley (1982), and Dupuy (1985). The only effort in this direction of which I'm aware is a fairly sketchy one by Poole (2003). Poole's broad context is small-unit light-infantry tactics as practiced by the high-tech US Army and Marines against low-tech infantry, guerrillas, and other paramilitary units. Poole gives two statistical analyses – the Korean War and the Tet Campaign in Vietnam.

Case: Korean War, 1950–3

The best available statistics on battle casualties during the Korean War reveal a much less lopsided total than generally believed. If we combine the North Korean and the Chinese casualties, we get somewhere between 1.3 million and upwards of 1.5 million total communist battle casualties. If we combine the US and UN casualties and those of their South Korean allies, total casualties are somewhere between 1.13 million and 1.54 million. In other words, at close to 1.5 million each, the two sides achieved virtual parity. Given the fact that the US/UN forces had air-sea dominance and fielded a more high-tech heavy infantry, how did the Chinese and North Korean light infantry manage to match casualties on a nearly one-to-one basis? Poole's plausible explanation is that it was because the Chinese and North Korean light infantry were much better trained in deceptive tactics, right down to the squad level.

Case: Tet Offensive, Vietnam, 1968

The best available statistics on battle casualties during the sixty days of the Tet Offensive during the Vietnam War reveal a similarly much less lopsided total than generally assumed (this account is based on Poole 2003, 41, 42–3, 348). Poole remarks that, unlike in Korea, "this time, the Communists were not only badly outgunned, but also badly outnumbered." The combined Viet Cong and North Vietnamese Army (VC/NVA) attacked during Tet with only 67,000 combat troops against the combined 1,179,000 US and Army of the Republic of Vietnam (ARVN) defenders. Casualties for the VC/NVA ran from 19,000 to 40,000, against 45,000 defenders (24,000 US and 21,000 ARVN) – a surprisingly low casualty ratio for attackers. Again Poole, a Vietnam platoon commander, attributes this outcome to the enemy's greatly superior use of deception from division down to squad level.

Poole concludes that "[i]n effect, 'low-tech' Eastern armies have had to capitalize on the field skills of the individual soldier." These skills, practiced and honed to a degree found among American troops only in Special Forces, include all the arts of deception – camouflage, invisible movement, stealthy approach, night operations, decoys, lures, and so on (Poole 2003, 43–4).

Small-scale deception operations

In this section, I examine the costs and benefits of two types of small-scale deception operations – sniping and ambushes.

Sniping

Sniping is the smallest-scale and stealthiest military operation. Typically involving either a single sharpshooter or a two-person team, sniping is generally well worth the special efforts required in training, equipment, and administration. In addition to the opportunity for intelligence gathering, sniping operations lower enemy morale, slow their advancing patrols, and regularly yield the highest kill ratios of all types of infantry combat.

Sniping may be the single most cost-effective type of military operation, and it's certainly the most cost-effective infantry operation. This is demonstrated dramatically by the following statistics (Sasser and Roberts 1990, 152):

- At the end of World War I (1917–18) on the Western Front, American troops fired an average of about 7,000 bullets for each German soldier killed.
- In World War II (1939–45), Allied infantry fired approximately 25,000 bullets for each enemy soldier killed.
- During the Korean War (1950–3), UN troops fired 50,000 rounds for each enemy soldier killed.
- In Vietnam (1955–75) American combat troops (with M-14s and later with rapid-fire M-16s) fired at least 200,000 shots per kill.

• By contrast, the average trained sniper requires only 1.3 bullets per kill – just slightly more than the sniper's proud motto of "One shot – one kill" (Sasser and Roberts 1990, 2).

Ambushes

The ambush is another type of smaller-scale military deception operation that can be cost-effective. It has proved itself from ancient times to the Iraqi insurgency that began in 2003. However, the ambush has been grossly underrepresented in systematic studies (British Col. Malleson's *Ambushes and Surprises* [1885] notwithstanding). There is no excuse for this omission. The subject is both important and easily researched. I discovered this in 2003–4 when I collected over fifty personal-experience examples from as many of my American and foreign Special Ops students at the Naval Postgraduate School. My necessarily tentative conclusion from such anecdotal evidence is that ambushes are not only usually cost-beneficial but almost always so. And, while my evidence is skimpy, the same seems to be true of counterambushes.

Zero percent solutions

I refer to deception as the "1 percent solution," although admittedly, as some of the above cases show, it sometimes costs more. A 5 percent diversion of a commander's strength might not deter him from employing deception, but 10 percent would surely give him pause, and rightly so. And, except in extreme peril, planners would and should have difficulty in justifying a 20 percent or 25 percent diversion. But an unchallengeable case would be made if the cost was zero. Is that possible? Rarely, but yes, as the two following cases show.

Case: General Funston ends an insurrection, Philippines, 1901

One consequence of the American victory over Spain in 1898 in the Spanish-American War was that the United States annexed the Philippines (this account is drawn from Seelinger 2004; Editors of the *Army Times* 1963, 23–30; and Funston 1911). The American occupation led the 30-year-old Chinese-Tagalog leader *Presidente* Emilio Aguinaldo to launch a full-scale insurrection against the occupiers on 4 February 1899. Two years and 6,000 American soldier casualties later, there seemed no end to the affair. Moreover, with the first flush of victory past, the American press and public had turned against the US occupation.

Then, on 8 February 1901, thirty-five-year-old Medal of Honor recipient Brig. Gen. of Volunteers Frederick Funston, a district CO at San Isidro in central Luzon, read a dispatch intercepted with one of Aguinaldo's couriers. It ordered insurgent General Lacuna to send 200 soldiers from his brigade to Aguinaldo's headquarters. The courier revealed that this HQ was currently encamped at Palanan, a small riverside village in a remote region of northern Luzon. He further disclosed that Aguinaldo's HQ guard was only fifty strong. Funston knew that the only approach was overland up a long tortuous road through a dense forest inhabited by rebel-friendly villagers. He realized that any direct assault would only run a gauntlet of ambushes, ending at a deserted rebel HQ. So Funston proposed a bold and clever deception to his commander in chief, Gen. Arthur MacArthur, who approved.

Funston formed a party of four American officers and 85 soldiers from the Macabebe Regiment. He'd chosen the Macabebes because they physically resembled Tagalogs. Funston and his American officers (all disguised as privates) would pose as POWs of his Macabebes. The Macabebes would themselves simulate the advance half of the Tagalog reinforcements expected by Aguinaldo. Three weeks were spent in planning and preparations. The Macabebes were coached in Tagalog dialogue and outfitted with captured weapons and clothing. Previously seized letterhead of General Lacuna would provide forged documents to smooth their passage through enemy territory. On 6 March, Funston and his small party, totaling eighty-nine men, set out for Aguinaldo's HQ at Palanan.

The trip required several further ruses, each improvised to overcome unexpected problems, particularly during the final approach, a 110-mile march through jungle that began on 14 March. Nine days later, on 23 March, Funston's party reached Aguinaldo's camp.

On their arrival, Funston sent two of his trusted Filipino officers, Segovia and Placido, ahead, and they were personally greeted by Aguinaldo. Then, on Segovia's unobtrusive signal, Funston's main force opened fire on fifty of the surprised Tagalog guards, killing two and scattering the rest. Hearing the nearby rifle shots, Aguinaldo assumed it was his own guard firing in the air to celebrate the arrival of reinforcements and stepped to the window of his bungalow, where he was seized by Placido. Two Tagalog officers were killed and two surrendered before the rest of the insurgent officers and troops fled. Funston graciously accepted the *Presidente*'s surrender, along with that of two of his chief officers.

Funston's daring ploy effectively ended the Philippine insurrection. Aguinaldo's final act as President of the Philippine Republic was to order all his followers to cease resistance. Except for the separate insurgency by the Muslim Moros in the south, his order was fully obeyed within three weeks. That was the great strategic benefit of Funston's small-unit tactical deception operation. The cost had been, in effect, zero. And, had the expedition failed, it would have meant, at worst, the lives of 89 Americans and their allies. The only real cost of this successful deception was in reputation – the American press generally sneered at General Funston for having used such un-American tactics.

Case: Dr. Jones versus Hitler's flying bombs

The single most cost-beneficial deception ploy of World War II was perhaps the successful diversion of Hitler's rocket bomb attacks on London in 1944 (this account is drawn from Jones 1978, 420–4, and map opposite 301; and Irving

1964, 250–8). The Luftwaffe V-1 campaign aimed 2,340 of the terrifying "Flying Bombs" toward the heart of the British capital. V-1 bombs killed 5,500 Londoners and seriously injured 16,000. There would likely have been as many as 50 percent more casualties if it hadn't been for the Germans' persistent misperception that most of the hits were overshooting their intended aiming point, Tower Bridge, at the geographical center of the great metropolis. In fact, most bombs were falling short, onto the much less densely populated southern suburbs.

To adjust range, the Germans depended on individual time-and-place bomb reports from their spies in London. Unknown to them, *all* their agents were under British control. Taking full advantage of this fact, Dr. R. V. Jones, a world-class physicist at the Air Ministry, conceived and designed a scheme of plausible agent dispatches that reported real hits but faked their times of impact. This elegant ruse induced the Germans to steadily readjust their real aiming point ever further short of their intended one. By the end of the campaign, the Flying Bombs had been lured four miles south of their original bull's-eye.

There are two circumstances that could have foiled Dr. Jones' scheme. First, the balance of correct reporting on location of hits with misreporting of their times was designed to seem consistent with any enemy aerial photoreconnaissance evidence. In fact, but unknown to the British at the time, the Luftwaffe had managed (accidentally) to put only one photo-recon aircraft over London during the entire ten weeks of the V-1 bombing campaign. Second, and also not suspected by the British, the Germans had placed radios on several of their Flying Bombs precisely to get a radio direction-finding fix on the time and place of impact. German air intelligence correctly saw that the combined plots showed that they had been aiming low. However, they concluded that, because their eye-witness human intelligence was reliable, it must be the radio direction-finding equipment that was inexplicably faulty!

Costs and benefits of counterdeception

It is harder to judge the relative cost of counterdeception – the detection of deception. But it's fair to say that in the majority of cases, most of the overall costs of detecting deception are prepaid by already existing intelligence capabilities. Indeed, as I've pointed out elsewhere, consistently effective deception analysis doesn't require more analysts or a bigger budget but simply different recruitment criteria, teaching tools, training procedures, and operational protocols (Whaley, *The Maverick Detective, Or The Detection of Deception* [unpublished draft, 2005]).

Conclusion and recommendations

Military leaders are understandably reluctant to accept any additional demands on their budgets and manpower that might detract from the combat effectiveness of their units. Moreover, they will rightly resist institutional additions to their already cluttered tables of organization and equipment. For example, the innovation in World War II of special psychological warfare units met widespread and serious resistance from puzzled field commanders who (justifiably) failed to see precisely what concrete military benefits balanced the all too apparent expense and inconvenience of their leaflet, loudspeaker, and propaganda radio units. Thus, military commanders need assurance that the investment cost of deception is low – both absolutely and relatively – and that the combat benefits are high.

The main conclusion of this chapter is simple. Deception is worth the cost and effort in almost all situations. So how can we improve our ability to practice deception and make our deceptions more effective, with more consistent chances of success?

First, we can profit from further systematic research to verify the findings I've presented here and to extend this research into other types of deception/counterdeception situations as well as more recent examples. I strongly recommend cost-benefit studies of three particularly rich data sources:

1   The National Training Center at Fort Irwin, where the US Army collects extensive and detailed data on real-time war games in which deception strategies have a prominent role.
2   After-battle studies from Somalia, the Balkans, Afghanistan, Iraq, and Lebanon, with a particular focus on ambush and counterambush actions.
3   Detailed interviews and oral histories from Special Ops veterans of the above five combat zones. These warriors are a rich source of anecdote. For example, from nearly 100 American and foreign Special Ops officers in 2003–4 from whom I obtained details of ambush situations, all but one were able to give revealing details. These men had been personally involved in ambushes, either as ambushers or ambushees, in combat or during training exercises.

Data of this type should be collected and collated to support analyses of asymmetric insurgency and terrorist operations. While I believe the broad outlines of deception cost-benefit findings presented in this chapter will be confirmed by further study, new analyses will flesh out and fill in the details. This will help us to better anticipate and master the challenges posed by adaptable and deceptive opponents. However, one of the most revealing points that emerges from this survey of case studies is that commanders need not wait for their think tanks to come up with solutions. Most of the various commanders seen in the cases I presented proved themselves sufficiently imaginative to solve a difficult problem. They found what I call the "third option," an unexpected solution to a complex problem. But then these were exceptionally qualified staff officers like Maj. Ingersoll, commanders like Maj. Funston, and scientists like Dr. Jones – all of whom possessed more than the usual "prepared mind" plus the imagination to create appropriate solutions.

Notes

1 Elsewhere I dubbed this the "Five Patton's Ruse," an error later corrected by reading the then recently declassified official history of the 23rd Headquarters Special Troops. In my interview with Ingersoll, he had exaggerated – as his deputy, Capt. Wentworth Eldredge, had warned me Ingersoll tended to do. See Whaley and Bell 1982, 379–81.
2 Price (1967, 243) estimates that "more than one thousand [RAF] bombers and their crews" were saved from December 1942 to October 1944 by electronic countermeasures alone. During that period, 4,490 RAF bombers were lost on night raids. Baxter (1948, 169) reports an official American estimate that "radar counter-measures saved the United States Strategic Air Force based in England alone 450 planes and 4,500 casualties."

Annotated bibliography

Well over a thousand books and articles have been devoted to military deception. Of these, however, those below marked with an asterisk (*) are the very few I've found that discuss, explicitly or implicitly, both the benefits and the costs of deception or include relevant numerical or statistical data.

*Alcala, Robert W., *Effective Operational Deception: Learning the Lessons of Midway and Desert Storm*, Newport, RI: Naval War College, February 1995 (22 pp).

Atkinson, Rick, *Crusade: The Untold Story of the Persian Gulf War*, Boston: Houghton Mifflin, 1993.

*Barkas, Geoffrey, *The Camouflage Story (from Aintree to Alamein)*, London: Cassell and Company, 1952.

*Barkas, Geoffrey, and John Hutton, "Camouflage of Airfields in the Middle East, 1941–42," *RAF Quarterly*, vol. 5, no. 2 (April 1953): 112–20.

*Barton, S. W. [pseudonym of Barton Whaley], and Lawrence M. Martin [pseudonym of Ladislav Bittman], "Public Diplomacy and the Soviet Invasion of Czechoslovakia in 1968," in Gregory Henderson (ed.), *Public Diplomacy and Political Change: Four Case Studies*, New York: Praeger, 1973, 241–314. On the Soviet deception operations that largely succeeded in concealing the Soviet intent to invade Czechoslovakia.

Bekker, Cajus, *The Luftwaffe War Diaries*, Garden City, NY: Doubleday, 1968.

Bell, J. Bowyer and Barton Whaley, *Cheating and Deception*, New York: St. Martin's Press, 1982.

Bittman, Ladislav, *The Deception Game*, Syracuse, NY: Syracuse University Research Corporation, 1972.

*Brugioni, Dino A., "If You Can't See It, You Can't Hit It," *Air Power History*, vol. 45, no. 4 (Winter 1998): 18–25. On the use of smoke to conceal tactical military operations.

*Bryant, Melrose M., *Deception in Warfare: Selected References from Air University Library Collection*, rev. edn, Maxwell Air Force Base, AL: Air University Library, May 1986.

Callwell, Sir C. E. (Major-General), *The Dardanelles*, London: Constable, 1919.

Churchill, Winston, *The Second World War*, vol. 4, Boston: Houghton Mifflin, 1950.

*Cohen, Fred, Dave Lambert, Charles Preston, Nina Berry, Corbin Stewart, and Eric Thomas, *A Framework for Deception*, Fred Cohen and Associates, 13 July 2001. Available online at all.net/journal/deception/Framework/Framework.html. Finds deception to be cost-effective for protecting information from computer hackers.

Copeland, Miles, *Without Cloak or Dagger*, New York: Simon & Schuster, 1974.

*Daniel, Donald C. and Katherine L. Herbig (eds), *Strategic Military Deception*, New York: Pergamon Press, 1982.

Davies, Merton E., and William R. Harris, "RAND's Role in the Evolution of Balloon and Satellite Observation Systems and Related US Space Technology," Santa Monica: RAND, September 1988 (RAND Report R-3692-RC).

*Despres, John, Lilita Dzirkals, and Barton Whaley, *Timely Lessons of History: The Manchurian Model for Soviet Strategy*, Santa Monica: RAND, July 1976 (RAND Report R-1825-NA). A pioneer analysis of the Soviet surprise-through-deception invasion of Manchuria in August 1945. See particularly Chapter 2 ("Strategic Design for Surprise Offensive"), 13–24, which was Whaley's main contribution to this paper.

*Dobinson, Colin, *Fields of Deception: Britain's Bombing Decoys of World War II*, London: Methuen, 2000.

Dupuy, Trevor, *Numbers, Predictions, and War*, Fairfax, VA: Hero Books, 1985.

Editors of the Army Times, *The Tangled Web*, Washington, DC: Luce, 1963.

*Eldridge, Justin L. C. (Captain), "The Myth of Army Tactical Deception," *Military Review*, vol. 70, no. 8 (August 1990): 67–78. A devastating critique of US Army tactical deception practice and doctrine.

*Eller, Joseph M., *Estimate of Cost of False Targets as Air Defense Auxiliaries*, Baltimore: Operations Research Office, Johns Hopkins University, May 1961 ("Confidential," declassified in 1970; published in fifty copies; 11 pp). This is not really a useful product. It gives the specs and estimated costs of seventeen types of simulated decoys (ranging from tanks to trucks) and the percentage of some types of decoy equipment to the real as used in certain US Army units, with no comparison of these costs with those of the real items, no assessment of effectiveness, and no conclusions.

Fall, Bernard B., *Hell in a Very Small Place: The Siege of Dien Bien Phu*, Philadelphia: Lippincott, 1967.

*Feer, Fred[ric] S., *Thinking-Red-in-Wargaming Workshop: Opportunities for Deception and Counterdeception in the Red Planning Process*, Santa Monica, CA: RAND, May 1989 (published as paper P-7510; 12 pp.). See also Eldridge 1990.

FM 90–2 (Tactical Deception), Washington, DC: Headquarters, Department of the Army, 2 August 1978. Particularly good in showing how high-quality operational deception can be had at little cost in diverting personnel and materiel.

FM 90–2 (Battlefield Deception), Washington, DC: Headquarters, Department of Army, 3 October 1988.

Funston, Frederick, *Memories of Two Wars*, New York: Scribner's Sons, 1911.

*Gawne, Jonathan, *Ghosts of the ETO: American Tactical Deception Units in the European Theater, 1944–1945*, Havertown, PA: Casemate, 2002 (342 pp). On the US Army's 23rd Headquarters Special Troops. The most detailed book on the subject, comprehensive except that the author doesn't look on the other side of the hill to discover how all this flim-flam was being perceived by the enemy.

*Gerard, Philip, *Secret Soldiers: The Story of World War II's Heroic Army of Deception*, New York: Dutton, 2002. A detailed unit history of the US Navy's development of the Beach Jumpers under Lt. Cmdr. Doug Fairbanks Jr and the Army's 23rd Headquarters Special Troops and 3133rd Sonic Deception Company.

*Greenberg, Irwin, "The Role of Deception in Decision Theory," *Journal of Conflict Resolution*, vol. 26, no. 1 (March 1982): 139–56. A pioneering effort to factor deception into game-decision theory.

*Greenberg, Irwin, "The Effect of Deception on Optimal Decisions," *Operations Research Letters*, vol. 1, no. 4 (September 1982): 144–7.

*Hartcup, Guy, *Camouflage: A History of Concealment and Deception in War*, London: David and Charles, 1979. The subtitle is grossly misleading. This book is narrowly limited to British military camouflage history. The author's timid excursions into Russian and Japanese camouflage are either speculative or wrong.

*Hauser, Marc D., "Costs of deception: Cheaters Are Punished in Rhesus Monkeys (*Macaca mulatta*)," *Proceedings of the National Academy of Sciences of the United States of America*, vol. 89, no. 24 (15 December 1992): 12137–9. Direct observation and field experiments both indicate that the rhesus monkeys on Cayo Santiago, Puerto Rico, who attempt to conceal (by failing to call) their discovery of food from their group members and are discovered with food are subject to more aggression. "Such costs may constrain the frequency with which deception occurs in this and other populations" (p. 12137).

*Haveles, Paul A. (Captain), "Deception Operations in Reforger 88," *Military Review*, vol. 70, no. 8 (August 1990): 35–41.

*Hewish, Mark and Bill Sweetman, "Hide and Seek: Camouflage, Concealment, and Deception: You Can't Attack What You Can't Find," *Jane's International Defense Review* (April 1997): 26–32.

Hoopes, Roy, *Ralph Ingersoll: A Biography*, New York: Atheneum, 1985.

Irving, David, *The Mare's Nest*, London: William Kimber, 1964. A detailed account of the German V-1 "Flying Bomb" and V-2 rocket development and operations during World War II and the British countermeasures. A fine job, as Dr. R. V. Jones later commented, before Irving drifted off into Holocaust denial.

*Jones, R[eginald] V[ictor], *Most Secret War*, London: Hamish Hamilton, 1978 (556 pp). Republished in the United States with slight changes as *The Wizard War: British Scientific Intelligence, 1939–1945*, New York: Coward, McCann, and Geoghegan, 1978. I cite from the US edition.

Keegan, John, *The Iraq War*, New York: Knopf, 2004.

*Kneece, Jack, *Ghost Army of World War II*, Gretna, LA: Pelican, 2001. On the US Army 23rd Headquarters Special Troops. Rough cost-benefit estimates are given on pp. 79–93, 280.

Liddell Hart, B. H., *Great Captains Unveiled*, London: Faber & Faber, 1929.

Liddell Hart, B. H., *Greater Than Napoleon*, London: Faber & Faber, 1941.

*Liddell Hart, B. H., *Strategy: The Indirect Approach*, New York: Praeger, 1954. The classic study of military surprise through deception. It is an enlarged, updated, and rewritten version of the author's brilliant *The Decisive Wars of History: A Study in Strategy* (London: Bell, 1929).

*Malleson, G[eorge] B[ruce] (Colonel), *Ambushes and Surprises; Being a Description of Some of the Most Famous Instances of the Leading into Ambush and the Surprise of Armies, from the Time of Hannibal to the Period of the Indian Mutiny*, London: W. H. Allen & Co., 1885.

*Mendelsohn, John (general editor), *From Normandy into the Reich*, New York: Garland Publishing, 1989 (4 files). US Army tactical deception operations on the Continent, 1944–5, three by the American 23rd Headquarters Special Troops and one by US Sixth Army Group's 2nd Tactical Headquarters "A" Force.

Middlebrook, Martin, *The Nuremberg Raid*, New York: Morrow, 1974.

Moorehead, Alan, *Gallipoli*, New York: Harper, 1956.

Official History 23rd Headquarters Special Troops, 1945 (129 pp). On the US Army tactical deception unit in Europe, 1944–5. Particularly interesting in that, without access to German records, the 23rd greatly *underestimated* its success.

*Photo Fakery: The History and Techniques of Photographic Deception and Manipulation, Dulles, VA: Brassey's, 1999.

*Poole, H. John, The Tiger's Way: A US Private's Best Chance for Survival, Emerald Isle, NC: Posterity Press, 2003. Adds Russian and German small unit tactics to his previous Asian model. Includes in "Appendix A: Casualty Comparisons" (pp. 347–8) some relevant statistics.

Possony, Stefan T., and J. E. Pournelle, The Strategy of Technology: Winning the Decisive War, Cambridge, MA: University Press of Cambridge, 1970.

*Price, Alfred, Instruments of Darkness, London: Kimber, 1967. There is a revised 1977 edition. On the aerial electronic hider-hunter game, where each new method of concealment provokes a new instrument for its detection and vice versa.

*Reed, George L. (Captain), "Voices in the Sand: Deception Operations at the NTC," Armor, vol. 97 (Sep.–Oct. 1988): 26–31. Detailed summary of the extensive but highly cost-effective use of deception by the OPFOR (then the 63rd Regiment of the First Armored Division) at the Fort Irwin (California) National Training Center. Capt. Reed was the 1–63 Armor S4 with the OPFOR at Fort Irwin at the time of writing.

Richardson, Lewis, Statistics of Deadly Quarrels, Pittsburgh, PA: Boxwood Press, 1960.

Ronaldshay, Earl of, The Life of Lord Curzon, vol. 3, London: Benn, 1928.

The Royal Air Force, vol. 3, London: Her Majesty's Stationery Office, 1954.

Sasser, Charles W. and Craig Roberts, One Shot – One Kill, New York: Pocket Books, 1990.

Seelinger, Mathew J., "A Desperate Undertaking: Funston Captures Aguinaldo," www.armyhistory.org, 2004.

*Sherwin, Ronald G. and Barton Whaley, "Understanding Strategic Deception: An Analysis of 93 Cases," in Donald C. Daniel and Katherine L. Herbig (eds), Strategic Military Deception, New York: Pergamon Press, 1982, 177–94. A secondary analysis of data on strategic-level deceptions in an updated version of Whaley (1969).

*Stanley, Roy M., II, To Fool a Glass Eye: Camouflage versus Photoreconnaissance in World War II, Washington, DC: Smithsonian Institution Press, 1998.

*Sykes, Steven, Deceivers Ever: Memoirs of a Camouflage Officer, 1939–1945, Tunbridge Sykes, Wells, Kent: Spellmount, 1990.

*Van Vleet, John Anton, Tactical Military Deception, Monterey, CA: Naval Postgraduate School, September 1985. Mainly a secondary analysis of the raw coded data on tactical cases in Whaley's Stratagem.

Verrier, Anthony, The Bomber Offensive, London: Bratsford, 1968.

Von Falkenhayn, [General] Erich, The German General Staff and Its Decisions, New York: Dodd, Mead, 1920.

Webster, Sir Charles and Noble Frankland, The Strategic Air Offensive Against Germany, 1939–1945, vol. 2, London: Her Majesty's Stationery Office, 1961.

*Whaley, Barton, Stratagem: Deception and Surprise in War, Cambridge, MA: Center for International Studies, MIT, 1969 (965 pp, multilithed, 150 copies).

Whaley, Barton, Codeword Barbarossa, Cambridge, MA: MIT Press, 1973.

*Whaley, Barton, "Toward a General Theory of Deception," Journal of Strategic Studies (London), vol. 5, no. 1 (March 1982): 179–93.

*Whaley, Barton, Covert German Rearmament, 1919–1939: Deception and Misperception, Frederick, MD: University Publications of America, 1984 (149 pp). A case study relevant to both deception and counterdeception, specifically arms control verification intelligence and order of battle-type intelligence. Written in 1979.

Whaley, Barton, Detecting Deception: A Bibliography of Counterdeception across Time, Cultures, and Disciplines, 2nd edn, Washington, DC: Foreign Denial and Deception

Committee, Office of the Director of National Intelligence, March 2006 (660 pp). A comprehensive and annotated bibliography of nearly 2,500 titles.

*Whaley, Barton, *et al.*, "Thoughts on the Cost-Effectiveness of Deception and Related Tactics in the Air War, 1939–1945," Research Paper, Office of Research and Development, Central Intelligence Agency, and Mathtech, Inc., Rosslyn, VA, March 1979 (170 pp). A case study of deception costs and benefits and the cycle of electronic countermeasures and counter-countermeasures employed by the British and German air forces in World War II.

*Wildhorn, Sorrell, *The Potentialities of Deception as a Survival Aid for a Retaliatory Missile Force*, Santa Monica, CA: RAND Corporation, 1962 (39 pp).

7 Strategy and psychological operations

Hy S. Rothstein

In early 2004, a Department of Defense Information Operations Steering Committee decided that a review of psychological operations (PSYOP) lessons learned from Operation Iraqi Freedom (OIF) was needed. The National Defense University published a report in September 2005 that reviewed PSYOP in OIF as well as in Operation Enduring Freedom (OEF). The report found that PSYOP lessons learned were consistent with historic findings, namely, that PSYOP continues to suffer from a lack of national-level themes, a slow or unresponsive product approval process, questionable product quality, and an overall lack of resources, including sufficient force structure.[1] Why do the same problems continue to undermine the effectiveness of PSYOP? Why is it that lessons learned seem more like lessons observed?

The National Defense University report deflects the real problem: PSYOP lacks acceptance and understanding within the contemporary national security structure. As a result, the role of PSYOP has not received its proper due despite notable but singular achievements in the field. The psychological dimension of war is an integral part of warfare, not a separate entity reserved for a few specialists who are tasked with adding some degree of value to a sterile military calculation of correlation of power. Getting an opponent commander to act in a manner advantageous to you and disadvantageous to him requires that he be made to think in a manner consistent with the desired action. This is the essence of strategy, and it is profoundly psychological. If PSYOP is integral to strategy, then strategic thinking within the US national security structure may be deficient.

In this chapter, I focus on the strategic level of war and attempt to accomplish four things. First, I review the conduct of PSYOP in OEF and OIF. Second, I look at a model formulated by the strategist André Beaufre that can point the way toward developing effective strategy. Third, using Beaufre's concepts, I briefly assess US activities in the "Global War on Terror" in two different countries, Iraq and the Philippines. I conclude the chapter with recommendations on improving the psychological dimension of strategy.

PSYOP in recent operations

Operation Enduring Freedom

Operation Enduring Freedom in Afghanistan required both large-scale combat operations to topple a regime and small unit operations to capture or kill terrorists. Additionally, stability operations were necessary to permit democratic elections to take place and to allow the new government in Kabul to establish itself. PSYOP involvement in OEF began immediately after the terrorist attacks of 11 September 2001. A tactical PSYOP detachment began an analysis of Afghanistan, its populace (including the Taliban), and al-Qaeda. In late September, US Central Command (CENTCOM) requested that US Special Operations Command activate a Joint Psychological Operations Task Force (JPOTF) to plan PSYOP in support of the military campaign. The JPOTF was placed under the operational control of CENTCOM with the mission of supporting CENTCOM as it carried out short-term strike operations, long-term counterterrorism operations, and engagement activities.[2]

The CENTCOM campaign plan provided guidance for conducting PSYOP in CENTCOM's area of responsibility. The plan defined PSYOP target audiences, objectives, and themes. The primary objectives were to:

- Shift the debate from Islam to terrorism and to counter adversarial propaganda.
- Discourage interference with humanitarian affairs activities.
- Support objectives against state and nonstate supporters and sponsors of terrorism.
- Disrupt support for and relationships within terrorist organizations.[3]

The PSYOP objectives were to isolate both the Taliban and al-Qaeda from domestic and international support while increasing the perceived legitimacy of US operations. The result would be a reduction of Taliban and al-Qaeda effectiveness by undermining their morale and willingness to fight.

Prehostility PSYOP actions focused on the development of leaflets and radio scripts. In the early stages of the operation, PSYOP was focused on product development and distribution. The JPOTF developed radio scripts that were given to the crews of specially outfitted EC-130 (Commando Solo) aircraft for broadcast over Afghanistan; broadcasting began on 5 October 2001, two days before the start of major combat operations. In Afghanistan under the Taliban, however, possession of a radio was a crime, and thus few were available. Consequently, small battery-powered transistor radios had to be distributed by airdrop and by tactical PSYOP teams operating with Special Forces detachments. Propaganda leaflets were shipped to PSYOP support elements in Diego Garcia to be assembled into leaflet bombs and dropped by B-52s. The first leaflets were dropped into Afghanistan on 14 October 2001, almost a week after combat operations began.[4]

PSYOP messages initially encouraged the Taliban to cease support of al-Qaeda and Taliban military forces to return to their homes. The messages demonstrated some sophistication in trying to expose members of al-Qaeda as foreign intruders who manipulated the Taliban. However, attempts to win over the Taliban were soon dropped, and the distinction between Taliban and al-Qaeda was eliminated. The PSYOP messages shifted to emphasizing the power and determination of the United States to punish and destroy the terrorists, and to calling on the Afghan people not to support al-Qaeda or the Taliban, but rather to join in a common cause with the coalition against those parties. The PSYOP products portrayed the Taliban as controlled by foreigners who had brought misery to Afghanistan and asserted that the coalition forces were there to help the Afghan people regain their self-determination. The operation put much effort into explaining why the United States and its coalition partners were in Afghanistan. Messages emphasized that the coalition was not there to provoke a conflict with Islam or to occupy Afghanistan, but rather to render it safe from terrorism and then depart.[5]

PSYOP produced far more public information than military messages. In an effort to build goodwill with the Afghan people, PSYOP messages focused on mine avoidance, the advantages of accepting vaccines for children, and the importance of using potable water. Most of these products could have been used anywhere, but some explicitly demonstrated sensitivity to Afghan culture by acknowledging an Islamic holiday and indicating that American food donations complied with Islamic dietary law. The PSYOP military messages emphasized US power (airpower in particular) and the inevitable destruction of all terrorists. Reflecting the assumption that terrorists could not be persuaded to abandon their cause (and also mirroring the US domestic mood), the messages did not request surrender but sought to demoralize enemy combatants by communicating our intent and ability to kill them.[6]

PSYOP units supported Special Operations forces with loudspeaker operations and the distribution of leaflets. The fall of Kabul, the spiritual home of the Taliban, led to the complete disintegration of Taliban military forces. The US military turned its attention to the hunt for Osama bin Laden and his al-Qaeda operatives. PSYOP supported these operations with products advertising a reward program for information on terrorists. As a by-product of their face-to-face interaction with Afghan villagers, members of PSYOP teams became a valuable source of knowledge about the human terrain at the village and tribal levels in Afghanistan.[7]

PSYOP units adapted to support postcombat operations. Significant PSYOP resources were committed to humanitarian aid efforts by giving supplies to local Afghan schools and distributing blankets and medicine to hospitals. Messages also emphasized that stability was key to the continued delivery of aid. PSYOP also supported the political process by communicating the need for a unified Afghanistan and portraying President Hamid Karzai as someone who would bring prosperity to the population rather than the horrors experienced during the rule of the Taliban.[8]

Operation Iraqi freedom

Although still classified, the PSYOP objectives for OIF may be easily surmised from the products that were prepared for operations. PSYOP objectives generally sought to isolate the Iraqi regime from domestic support and to reduce the effectiveness of Iraqi forces by undermining their morale and willingness to perform their missions.

Combat operations for OIF began on 21 March 2003. PSYOP planning was facilitated by much of the support, command, and production capability that was already in place as a result of OEF. PSYOP planners also benefited from extensive experience resulting from many years of conducting operations against Iraq in support of Operations Southern Watch, Northern Watch, and Desert Fox.[9]

Planning for PSYOP against Iraq officially began on 15 November 2002. In January 2003, the JPOTF commander and staff moved from CENTCOM headquarters in Tampa, Florida, to Kuwait to better support preparations for OIF. In February 2003, the JPOTF commander and three staff members relocated to Qatar along with CENTCOM headquarters. PSYOP planning for OIF actually started almost a year before the start of major combat operations. The total number of PSYOP personnel committed to the JPOTF in support of OIF fluctuated between 600 and 700, although only a handful of these personnel were forward-deployed with the JPOTF commander. PSYOP forces were tasked with several missions, including:

*   Establishing an audience for radio broadcasts in Iraq.
*   Undermining the confidence of the Iraqi military and Iraqi security forces in Saddam Hussein's regime.
*   Degrading Saddam Hussein's confidence in his ability to control Iraq.
*   Deterring the use of weapons of mass destruction (WMD).
*   Assisting in locating enemy forces and weapons.
*   Gaining widespread support for coalition forces.
*   Informing Iraqi civilians that coalition forces would not do them harm.
*   Distributing radios to the civilian population so that they could listen to coalition radio broadcasts.[10]

These missions were accomplished with extensive use of radio broadcasts and leaflet dissemination. Broadcasts into Iraq originated from Kuwait, from aerial broadcasts from Qatar-based Commando Solo aircraft, or from maritime broadcasts from ships in the Arabian Gulf. PSYOP teams also transmitted radio programs via the Special Operations Media System-B (a mobile ground-based radio and television broadcasting station). Some PSYOP messages specifically attempted to deter Iraqis from repairing communications infrastructure that was being bombed and warned Iraqi air defense units not to target or shoot at US planes. The radio messages emphasized the legitimacy of UN votes to sanction compliance with resolutions and the US military buildup to enforce compliance if necessary. Leaflet drops and radio broadcasts continued throughout the months before the start of combat operations. The prehostilities messages emphasized the inevitability of coalition success and the dangerous and counter-productive nature of mining waterways, sabotaging oil installations, blowing up dams, and using WMD.[11]

Coalition forces entered Iraq on 21 March 2003. At this point, most leaflets were directed to Iraqi military units. Leaflets were designed to convince Iraqi forces to surrender or desert. About half of the leaflet messages prepared for this purpose were generic, and the remainder specifically targeted Iraqi units that failed to comply with coalition instructions. The message to Iraqi soldiers was clear: Go home where you are needed, or be totally destroyed, as were those units that resisted. The leaflets encouraged the Iraqi soldier not to sacrifice his life for a bad regime. The messages were sympathetic to the common Iraqi soldier and communicated a preference for not having to kill him.[12]

Messages were also directed to Iraqi civilians, who were asked to avoid combat areas and to listen to the coalition radio broadcasts. All messages emphasized the incompetence, corruption, and illegitimacy of Saddam Hussein and his lieutenants. Messages directed to the general public emphasized political themes, whereas those directed to military audiences emphasized the futility of resisting the coalition.[13]

Tactical PSYOP teams were attached to ground forces and played an important support role for infantry and Special Operations units. PSYOP teams supporting ground forces communicated surrender appeals to enemy units via loudspeakers. They also provided information to civilians and requested noninterference with combat operations in order to avoid noncombatant casualties. An example of the value of loudspeaker operations, from operations that occurred on 25 March at an-Nasiriyah, was recorded by the Center for Army Lessons Learned:

> TPT [tactical PSYOP team] 1141 was supporting Task Force Tarawa assigned to the 1st Marine Expeditionary Force (I MEF). Tarawa was fight-ing paramilitary forces that threatened to bog down the Marines' advance. Iraqi paramilitary forces hiding in a hospital were sniping and firing mortars and machine guns at Marines crossing the bridge over the Euphrates. TPT 1141 broadcast a surrender appeal and a statement about the inevitability of their defeat, and told [them that] we would drop bombs and artillery on the hospital if they did not surrender. Approximately 10 minutes into the broad-cast, Iraqi personnel emerged from the hospital and complied with TPT 1141's instructions.[14]

TPT 1141 also supported Task Force Tarawa by ensuring safe passage of civil-ians and by obtaining valuable intelligence from civilian sources in the process.[15]

Up to this point in the war, PSYOP products were direct, simple, and to the point. They focused on legitimizing coalition operations, delegitimizing Saddam Hussein, and providing public service messages to minimize unnecessary civil-ian casualties and to emphasize the need to support the new government.

PSYOP performance during stability operations offers an interesting contrast to PSYOP during combat operations. Allied PSYOP experts noted that US PSYOP forces were slow to transition from major combat operations to stability operations. This is not surprising, since US doctrine and associated tactics, techniques, and procedures do not clearly identify the difference between the PSYOP objectives for major combat operations and those for stability operations. The broad categories of PSYOP missions identified in doctrine cover both combat and stability operations; they include:

- Isolating an adversary from domestic and international support.
- Reducing the effectiveness of an adversary's forces.
- Deterring escalation by adversarial leadership.
- Minimizing collateral damage.[16]

However, at the tactical level, there are nontrivial differences in specific PSYOP missions. Arguably, the contribution of PSYOP is more critical during stability operations than it is during major combat operations. For example, since the focus in stability operations is to drive a wedge between irregular forces and their base of popular support, PSYOP must do more to reach the general populace. During major combat operations, PSYOP can afford to concentrate more on enemy combatants. Stability operations require that PSYOP forces do more to control the information available to the local populace. However, in most instances, it will not be possible to monopolize information channels or control their content. Under these circumstances, it is essential that US forces make a positive impression on the local populace through their words and actions.

Soon after major combat operations ended, it became evident that US forces were engaged in an increasingly difficult stability operation that required a course correction for PSYOP forces. On 17 July 2003, the JPOTF received a revised mission. The task force was to:

> Conduct PSYOP in Iraq to provide a secure and stable environment and facilitate development of a functioning civil administration, which adheres to the rule of law, promotes regional stability, and eliminates threats from terrorism and WMD. On order, transition responsibility to the Coalition Provisional Authority enabling the Coalition Forces Command forces to withdraw.[17]

Themes to support the revised mission were developed, and messages were disseminated primarily through face-to-face interactions and loudspeaker operations. PSYOP teams also accompanied infantry units in support of search operations. Through an interpreter, PSYOP teams read messages to the people occupying the buildings being searched to explain the purpose of the search operation, the intention of the search teams, and what was to be expected of the occupants once they vacated the buildings.[18]

It quickly became apparent that support for stability operations would require resources well beyond those normally available to PSYOP forces. Conducting successful PSYOP in an environment in which multiple competing sources of information are available to target audiences required more than standard PSYOP forces could deliver. During combat, the focus is usually on enemy forces that had limited sources of information. In stability operations, the target audiences are among the general population, where the competition for public influence can be fierce.[19]

Successful PSYOP is never guaranteed, and it wouldn't be even if we were able to dominate or compete fairly with all information sources available to a target audience. Words must be followed by supporting actions. The behavior of US forces can send powerful messages to the local population. Properly conceived and executed, tactical operations can help raise popular opinion about the US presence and purposes and ultimately reduce the cost of winning. This old lesson is being relearned by some US commanders, as the following excerpt from an internal Army report on the operation indicates:

> Current Iraqi sentiment has evolved from personal relationships between coalition soldiers and Iraqi citizens. Because these relationships differ from location to location and person to person, it is hard to correlate events and actions with relationship successes or failures. An armor[ed] task force commander in Baghdad described his methodology as that of plotting and measuring everything: "After a positive or negative event, he would have his staff evaluate all actions they had conducted before, during, and after the event. This would allow him to correlate activities with outcomes and develop TTP [tactics, techniques, and procedures] for future success." While not all actions provided equal measures or correlating events, this methodological approach helped establish a baseline of comparative success for his task force. It allowed him to utilize the Army approach of BDA [battle damage assessment] in regard to stabilization and support operations. This task force was very in tune with the local Iraqi populace in their area of operation. A bond was built and cultivated over many months, resulting in a trust between the military and civilian population. As a result of this established relationship, a new anxiety has developed due to the impending transfer of authority. The local leaders inquired about the replacement force and their capability to "be as good" as the current command.[20]

This anecdote demonstrates that tactical commanders can assess a target audience and effectively conduct PSYOP. Unfortunately, as the Army report notes, this commander's experience was not the norm:

> Some US units understand and use the concepts of setting objectives, developing themes, and setting measures of effectiveness. Others do not understand the process and therefore are just conducting operations without any measure of success or failure. Some commanders use tactical psychological

operations teams as a reactive measure when negative second or third order effects occur. Most IO [information operations] battle drills are reactionary in nature.[21]

Assessment

A complete assessment of PSYOP effects during OEF and OIF is difficult to undertake. In general, it is easier to assess tactical effects where specific behavior is requested, such as in surrender and desertion appeals or weapons buyback programs. On the other hand, efforts to change general target audience attitudes through radio broadcasts and other products are difficult to assess without a rigorous program to conduct surveys. One could say that tactical-level PSYOP is a more comfortable domain to work in than the operational domain, where anecdotal evidence, informal focus groups, and other media tend to provide indications of PSYOP effectiveness. However, it seems clear that PSYOP units supporting combat operations were better prepared to influence target audiences than were those supporting stability operations. Also, supporting tactical maneuver elements seemed to be favored over efforts that could affect the entire theater. This is ironic, considering that the role of PSYOP is generally considered more important for stability operations.

Perhaps the most salient point about the importance senior military leaders place on PSYOP is contained in the National Defense University report:

> Below the level of combatant commander, flag officer attitudes about the ability of PSYOP to create effects remain mixed. It is not surprising that in combat, many commanders will place greater confidence in kinetic weapons with which they are more familiar and which have more easily demonstrated effects. For many, substituting kinetic options with PSYOP products amounts to targeting on faith, since their actual effects are so difficult to observe and quantify.[22]

The impact of PSYOP activities will always be more obscure than that of kinetic weapons, but much more could be done to systematically assess PSYOP effects through dedicated intelligence support and interrogation of target audiences.

The National Defense University report emphasizes the more tangible return on investment for tactical PSYOP versus theater PSYOP. At the same time, the report acknowledges that PSYOP is more critical to the success of stability operations than it is to tactical operations. The complexities and resources needed to fix PSYOP for stability operations greatly exceed what is needed to improve tactical PSYOP support. Given limited resources and the general uneasiness expressed by senior leaders about the effectiveness of theater PSYOP, the report seems to prioritize improving the tactical-level over the theater-level PSYOP. This keeps PSYOP where it has always been – off in an obscure corner separated from war planners.[23]

PSYOP has an important force multiplier effect by helping to reduce the effectiveness of enemy forces and to deter escalation. Tactical PSYOP can be accomplished effectively and at low cost. Arguably, PSYOP at the operational and strategic levels could be left to the realm of public diplomacy, where integration of national themes and messages can best be achieved. Doing so would not have a harmful impact on tactical combat operations. However, while a well-executed PSYOP program that supports tactical combat operations can be expected to produce significant results, it is not as likely to produce long-term policy outcomes without leveraging coercion, without well-orchestrated and complementary public diplomacy and public affairs campaigns, and while it is in competition with numerous sources that are more credible to the target audiences.[24]

The information age

Operations in Iraq in 1991, the collapse of Yugoslavia, North Korea's aggressive behavior, and the humanitarian concerns and chaos produced by failed states such as Haiti and Somalia imply a variety of threats, some old and others new. Current political and military efforts in Iraq and Afghanistan dominate the agenda of America's decision makers, but it is unclear whether these efforts are episodic or a window into the future. Despite the wide variety of possible threats and actions, the notion of warfare entering the information age after the Cold War is almost universally accepted.

Although the information age has produced the concept of information operations, many of the components of information operations, and of PSYOP in particular, are not new.[25] The very nature of PSYOP is problematic for most Americans, as was clearly illustrated in the creation and quick demise of the Office of Strategic Influence or (OSI). OSI was created by the Department of Defense in October 2001 to support the war on terrorism through PSYOP in targeted countries. The office was intended to be a center for the creation of propaganda materials to mislead enemy forces and influence foreign civilian populations. Intense discussions on the purpose and scope of the OSI ensued in mid-February 2002, only after its existence became publicly known when information about its creation spread through US and foreign media. The discussions culminated in a public statement by Secretary of Defense Donald H. Rumsfeld in late February that the office would be closed down.[26]

More recently, in December 2005, US efforts to inform and influence the Iraqi population unraveled when it was revealed that US funds were being used to pay for favorable treatment in the Iraqi press. The general feeling was that this deliberate undermining of free inquiry in Iraq was morally troubling and counterproductive. Again, once the practice was exposed, the Pentagon announced that it would stop paying Iraqi journalists for favorable treatment.[27]

This is an unfortunate circumstance. It should not be hard to support aggressive clandestine action for developing democracy in Iraq. The United States ran enormous covert and overt operations throughout the Cold War. With the CIA usually in the lead, Washington spent hundreds of millions of dollars on book publishing, magazines, newspapers, radio stations, union organizing, women's and youth groups, scholarships, academic foundations, intellectual salons and societies, and direct cash payments to scholars, intellectuals, and journalists who believed in ideas that America thought worthy of support.[28]

For instance, it would be difficult, in an objective historical assessment of Radio Free Europe–Radio Liberty (RFE–RL), when it was still controlled by the CIA, to suggest that it was less truthful or more subject to political manipulation than today's Radio Liberty, which operates under the oversight of the independent Board of Broadcasting Governors. RFE–RL was one of the most successful "soft power" expenditures Washington ever made. Listeners were not concerned that the CIA backed the broadcasts. The issue back then, as it is today with Muslims reading or listening to US-supported material, is whether the content echoes the reality that they know.[29]

So where does all this leave us? One thing is clear: the war in Iraq has taken longer and cost more than was forecast, and the end is nowhere in sight. Consider the geopolitical setting four years after the Japanese attack on Pearl Harbor and compare it with the situation the United States now finds itself in, five years after the attacks on the World Trade Center and the Pentagon. Why have we been unable to find an effective strategy to defeat our enemies today? Perhaps America's success in the past was the result of something other than carefully developed strategy. There is good reason to question the ability of the United States to develop sound strategy for the successful pursuit of policy objectives. One model that illustrates this point and outlines a remedy is that proffered by the André Beaufre.

André Beaufre

Beaufre served on the French general staff from the mid-1930s to the mid-1960s. He had major responsibilities in World War II, in Algeria, in the Suez expedition, and in the North Atlantic Treaty Organization (NATO). As a result, he personally experienced many changes in warfare. He recorded his observations in several books, the best known being *Introduction to Strategy*.[30]

In *Introduction to Strategy*, Beaufre presents an interesting analysis of the evolution of warfare. Battles were preceded by a preparatory phase designed to pin down the enemy force, shake morale by means of fear, fatigue, and losses, and then concentrate forces against the decisive point on a flank or in the center. The preparatory maneuver was designed to cause the enemy to spend reserves either by committing them in false directions or squandering them in local actions. This pattern formed the basic principles that Beaufre used.[31]

For Beaufre, victory results from a successful preparatory maneuver that allows the culminating decisive attack, a situation he termed "freedom of action." Since blocking the enemy's preparatory maneuver was required, forces had to be divided for blocking, preparatory, and decisive actions, ideally using an optimal allocation or "economy of force." Beaufre's abstract formula for victory was to reach the decisive point by way of the freedom of action gained by sound economy of force.[32] Beaufre recognized and emphasized the psychological aspects of warfare throughout his work. In addition to material factors, the art of battle consists of maintaining and strengthening the psychological cohesion of one's own troops while at the same time disrupting that of the enemy's. The psychological factor is therefore key.[33]

Beaufre's analysis of combat operations provides a framework for the use of the military in pursuit of national objectives. Beaufre emphasizes the integration of politics and the military even more than Clausewitz. He insists on the concept of "total strategy," incorporating all instruments of statecraft. Total strategy is defined as the art of applying force so that it makes the most effective contribution toward achieving the ends set by political policy.[34]

All strategy for Beaufre boils down to an attempt to maintain one's own freedom of action while restricting that of one's opponent. Applied in an indirect manner, strategy usually involves the coordination of an exterior maneuver designed to fix or restrict the enemy, with an interior maneuver used to achieve an objective. Exterior maneuvers are applied on a worldwide scale and serve to restrict the freedom of action of the opponent through psychological, economic, or diplomatic means. The exterior maneuver prepares the way for the interior maneuver, which may be one of two types. The first involves a series of very rapid actions to gain limited objectives, interspersed with negotiations. The second is a protracted conflict that counts on erosion to weaken the enemy; it depends heavily on relative material and psychological strength.[35]

Beaufre notes that an army with superior resources and adequate striking power will act offensively, aiming for a Clausewitzian decisive battle; this is an offensive strategy using a direct approach. If superiority is less clear, if the objective is less than critical, if freedom of action is limited, or if direct offensive action simply seems less likely to succeed, two alternatives are suggested: either wearing down the enemy by defensive action followed by a counteroffensive or throwing the enemy off balance by a diversionary action prior to the real action. Finally, if military resources are inadequate or freedom of action is limited, military action will have only an auxiliary role. Essentially, marginalized military assets necessitates an indirect strategy that relies on psychological, diplomatic, and economic actions.[36]

Patterns of strategy

Strategic plans can generally be classified into a number of differing patterns depending on the importance of the issue at stake, the relative resources available to each side, and the freedom of action available to each side. The relationship among these three variables is the foundation of any strategy and requires a keen understanding of the geopolitical and military situations. The following are the most typical patterns according to Beaufre. If the objective is important and the resources available are large, a direct threat is likely to produce the desired political outcome. Alternatively, if the objective is important but the resources available are inadequate to exert a direct threat, an attempt to achieve the desired political outcome should be pursued by more subtle methods, which may be psychological, diplomatic, or economic.[37]

Next, the relative amount of freedom of action must be considered. If freedom of action is restricted and the resources available limited but the objective remains important, the pursuit of political objectives should be made by a series of successive actions in which a direct threat and indirect pressure are combined. If freedom of action is great but the resources available are inadequate to secure a military decision, a protracted strategy, the object of which is to wear down the enemy's morale and tire him out, is needed.

Beaufre does not specifically address a situation with restricted freedom of action, adequate resources, and an objective of great importance. Therefore, I add to his model the proposition that any significant restriction to freedom of action will likely necessitate an indirect strategy. The judicious use of resources is key to ensuring that the struggle can be maintained over a long period. At the same time, friendly actions must force the enemy to deploy an effort so great that he cannot maintain it indefinitely.[38]

Beaufre cautions us about the limitations of violent, direct conflict aimed at defeating an opponent's military forces in classic Clausewitzian fashion. Such operations are feasible when military capabilities are sufficient to produce both rapid and complete victory. If one does not have superior capabilities, the result may be a "prolonged period of mutual attrition out of all proportion to the issue at stake, at the conclusion of which both victor and vanquished emerge from the conflict completely exhausted."[39]

Conclusions

The patterns of strategy described above should be considered only as examples, not a complete typology. They are instructive in identifying the possibilities from which strategy has to choose and the nature of strategic thinking. The examples also show how other strategists have adhered unnecessarily to a predetermined path in prescribing a single form of strategy. Beaufre's own words best describe the dangers resulting from not adequately appreciating the factors that should determine strategy:

> Without adequate analysis of the factors governing strategy, the choice of a course of action has all too often been made out of habit or following the fashion of the moment. As a result, governments have not been in control of events, and clashes of purpose have led to fearful international catastrophes. The world of today is passing through an unparalleled crisis of readjustment, and at the same time science, industrialization and psychological action are making an increasing impact on the military art. More than ever before, therefore, is it vital that we should develop a method of thinking which will enable us to control, rather than be at the mercy of, events. That is why strategy is of such importance and such a problem of the moment.[40]

The United States and the global war on terror

Iraq: shock, awe, and transformation

Many, including President Bush and former Secretary of Defense Rumsfeld, have viewed OIF as the "new American way of war." To a large extent, it more closely resembles the way of the pre-Cold War era than a revolutionized way of warfare. Indeed, the attitude for dealing with Iraq inherent in this view seems to be rooted in the relative ease with which Americans were able to tame the frontier. Further-more, the new American way of war that results in "catastrophic success" simpli-fies war into a targeting drill. The enemy is a targeting array that once hit by precision munitions will surrender, and American goals will be achieved.[41]

The other similarity between OIF and America's past wars is the extent to which American military efforts barely considered political objectives. What proved to be the most important goal of the war – the political transformation of the Middle East – never defined the military mission in Iraq. Instead, the tradi-tional American way of war seems to characterize OIF. Specifically, it appears that the Pentagon's defense planners did not discover the real meaning of "regime change" until months after the initial combat operations began. For them, Operation Desert Storm in 1991 never quite ended. Thus, for the Penta-gon, the choices of military action were to contain Saddam's regime or launch a replay of Desert Storm designed to end it. The significance of the larger purpose of the war, political transformation of the Middle East, seems to have been lost on military planners.[42]

For many, OIF was a testing ground for the new warfare. This was especially true for both President Bush and Secretary Rumsfeld. Candidate Bush spoke about his vision of a military transformation that would rely on information technology and long-range, precision strike weapons systems. Secretary Rums-feld viewed military power as increasingly defined by mobility and swiftness and not by mass or size. Both initially opposed using the military for anything that resembled nation building. To help transform the Defense Department to operate under this new conception, Rumsfeld established the Office of Force Transformation in October 2001. The office's first director was Adm. Arthur K. Cebrowski (Ret.). Cebrowski was the ideal man to implement the national command authorities' vision of future warfare, since it was he, while on active duty, who helped develop and publicize a distinct form of warfare called network-centric warfare (NCW).[43]

NCW was the incarnation of what the president and defense secretary were looking for. In NCW, information superiority is exploited to generate increased combat power by linking sensors and decision makers with shooters to achieve greater lethality, survivability, and battlefield synchronization. The notion of knowledgeable entities generating increased combat power hangs on the thread of near-perfect intelligence. Knowing more about ourselves, the enemy, and the battlefield than the enemy knows, and keeping him from knowing about us, promises new capabilities in warfare. The proponents of NCW claim that it is revolutionary and transformative and that it provides the United States with the potential to go beyond attrition to a strategy based on "shock and awe." The concept of shock and awe is a logical byproduct of NCW. Both concepts rely on near-perfect intelligence to deliver long-range precision-guided munitions to destroy critical enemy systems and infrastructure. At the same time, these concepts minimize the friendly "footprint" in the theater.[44]

Shock and awe clearly reflects the psychological dimension of warfare. However, the concept has at least two problems that immediately come to mind. First, any honest appraisal reveals that shock and awe is as closely linked to Giulio Douhet's ideas about the relevance of air power theory as it is to NCW. Shock and awe rests on the belief that inflicting high costs quickly and with precision can shatter morale and undermine the basis for resistance. Unfortunately, there are sufficient data to show that airpower is neither a magic bullet nor a way to win inexpensively.[45] Additionally, any concept designed to have a psychological impact must take account of the fact that the target audience sets the parameters for influence, not the originator of the appeal. Predicting psychological effects is significantly more difficult than predicting physical effects.

The second problem of shock and awe (and NCW), and the most important one in the context of OIF, is that it does not translate the destruction of the enemy's ability to fight into the broader objective of regime change. One could argue that destruction of the opponent's combat capability must be complete before moving on to the next step. However, although the steps might need to be carried out sequentially, the planning should be concurrent. To illustrate, the application of shock and awe in Iraq used just enough troops to take the country but not enough to secure it. The use of massing effects rather than forces, thereby minimizing the theater footprint, is a key feature of this new-age warfare. This dynamic will continue to deliver solutions that leave people and equipment behind. Unfortunately, history clearly teaches us that it is much easier to destroy an existing regime than it is to establish a new stable one. Cycles of violence in Africa and Latin America, the Soviet failure in Afghanistan, and Napoleon's defeat in Spain demonstrate how difficult it can be to end the chaos that often follows the overthrow of even a relatively stable and secure government. Avoiding or reversing chaos and securing the support of the defeated populace for the new government require lots of boots on the ground. NCW has not liberated us from the need to occupy the terrain to win the peace.[46]

Thus, US war planners either woefully underestimated the number of troops needed to facilitate regime change or simply considered their task completed when the Iraqi army disintegrated. The Pentagon's plan did not reflect the president's policy objective of regime change. This is poor strategy. Secretary Rumsfeld's comment about the initial lawlessness that prevailed in Iraq after Saddam Hussein fell either acknowledges the planning shortfalls or illustrates his own arrogance. "Stuff happens and it's untidy, and freedom's untidy, and free people are free to make mistakes and commit crimes and do bad things."[47] Shock and awe and NCW have so far proved to be inadequate in delivering desired policy outcomes.

Separating war and politics

A detailed analysis of the evolution of US strategy in Iraq through the framework developed by Beaufre, although it would be useful, is beyond the scope of this chapter. Nevertheless, a simplified assessment is possible using Beaufre's variables of the importance of the issue at stake, the relative resources available to each side, and the freedom of action available to each side. This analysis can be further refined by including what Beaufre refers to as the "preparatory maneuver" that should lead to successful, culminating "decisive actions." The purpose here is not to review the material factors relevant to the art of war but rather the psychological cohesion of friend and fence sitter, while at the same time disrupting the psychological cohesion of the enemy. While it is possible to identify several criteria to measure progress, several points seem to be most relevant to influencing US freedom of action in OIF. Avoid the following:

1   Losing control of the countryside.
2   Alienating the general population enough to generate recruits for the enemy's cause.
3   Undermining your occupation.
4   Being forced to extend the length of the occupation beyond a period that is acceptable to your own population.[48]

Also, the high importance of the issues at stake in the Global War on Terror has been clearly expressed by the president and in various official documents. Finally, the relative availability of resources is assumed to favor the United States. Therefore, these two variables may be considered constant, so the focus is centered on relative freedom of action.

Anticipating and controlling events rather than being at their mercy is a fundamental aspect of strategy. The purpose of Beaufre's concept of preparatory maneuver is to a large extent aimed at this end. Successfully anticipating and controlling events will lead to increased freedom of action for friendly forces and decreased freedom of action for enemy forces. This is especially important in protracted war. But developing a strategy that results in increased freedom of action must begin with an understanding of the war aims.

Unlike the first US war against Iraq, which was waged for the limited objective of liberating Kuwait, the second US war was total war in which the United States sought to overthrow the Iraqi regime in order to disarm Iraq. It was the first time since World War II that the United States waged total war. It was a return to the tradition of Appomattox Court House, the Reims schoolhouse, and the deck of the USS *Missouri*. The numerous policy objectives delineated for the war can be reduced to two:

•   Prevent Saddam Hussein from acquiring WMD by removing him from power, destroying Iraqi WMD and their associated infrastructure, and establishing legitimate governance in Iraq that is at least benign toward the West.

- Transform the Middle East through the democratization of Iraq.[49]

The military defeat of Iraq and the overthrow of Saddam Hussein were never in doubt. Military experts, both optimists and pessimists, argued about the cost of victory but in no way clashed over the eventual military outcome. The US-dominated coalition forces seized Baghdad after only twenty-one days of combat operations. Iraq's best units proved to be no more than bomb bait for American airpower. The challenge, however, was not winning the war but winning the peace. Developing the means for exploiting military success was neglected in the "preparatory maneuver" phase.[50]

The United States had just demonstrated in Afghanistan the capacity to use force decisively. Many key officials in the Bush administration were critical of President Clinton for not using force decisively and constantly searching for approval from international institutions to legitimize military action. Furthermore, many felt that the Defense Department continued to be afflicted by the "Vietnam syndrome," in which war was thought to be warranted only when vital interests were threatened and military victory was guaranteed – principles that were codified in the Weinberger and Powell doctrines. Accordingly, an implied message in America's policy aims in Iraq was to demonstrate to the world that the United States would no longer permit its freedom of action to be constrained by a perceived need for international legitimization in the form of UN or NATO approval. The United States would act unilaterally regardless of world opinion.[51]

US preparatory actions resulted in constraining long-term freedom of action. It subordinated policy objectives to the narrow military objective of defeating Iraq and deposing Saddam Hussein. The preparatory maneuver generated the unfortunate strategic circumstance that resulted in the war's lacking international political legitimacy and the support of key allies, including Iraq's major neighbors. Near-unilateral military action in defiance of the UN and key NATO allies alienated friends and allies, especially in the Muslim world. Despite administration claims that OIF included a broad "coalition of the willing," only the United States and Great Britain contributed significant combat forces for the war. The rest of the coalition consisted of nominal political supporters. As a result, the war appeared to be an Anglo-American attack on an Arab state. To make things worse, a broad "coalition of the unwilling" believed that Iraq did not pose an imminent threat to the West and questioned the need to invade the country. The war thus underscored centuries of Western humiliation of the Arab world and was bound to have a negative impact on the achievement America's policy goals.[52]

Direct versus indirect approach and responding to catastrophic
success

Although alienating a significant portion of the international community had its costs, going forward with a coalition of the willing also had advantages. In the short term, freedom of action would clearly have been reduced if the United

States had been willing to accommodate the concerns of more key players in the international community. However, the United States wanted to maximize its freedom of action, and that is precisely why it sought a coalition of like-minded states.

Unfortunately, decisions designed to remove any uncomfortable operational constrains from the Anglo-American coalition's initial war plan undermined the coalition's freedom of action in the longer term and benefited future opponents. The preparatory phase consists of exterior and interior maneuvers. The exterior is intended to be applied on a worldwide scale and is designed to fix or restrict the enemy through psychological, economic, or diplomatic means. Properly executed, the exterior maneuvers will prepare the way for successful interior actions either to rapidly gain limited objectives or, in a protracted conflict, to erode the strength and base of support of the enemy. The policy goal of regime change meant total war. Therefore, any strategy had to consider the long-term dynamic nature of the environment. Although opting for a coalition of the willing created a short-term tactical advantage, it failed to consider the full meaning of the policy goal. The result has been an alienated international community unwilling to adequately support efforts to stabilize Iraq and an emboldened network of insurgents who are constantly increasing the costs for the dwindling coalition of the willing.

One can easily argue that the preparatory phase involved decisions and actions removed from military planners. Still, senior military leaders have the responsibility to ensure that military success results in successful political outcomes. More than three years into OIF, these outcomes seem far off. The great freedom of action enjoyed by US forces in the initial weeks of the campaign was the result of two factors – the almost unilateral nature of the war and the initial conventional disposition of the Iraqi military. The Iraqi military was arrayed in a manner that optimized the capabilities of US advanced weapons systems. In other words, the US military had a high degree of freedom of action to defeat the Iraqi military, a condition that also facilitated a direct approach to war. However, the rapid collapse of the Iraqi armed forces led to conditions that favored guerrilla warfare and insurgency for the enemy. This now meant that the relative advantage coalition forces had against the formerly conventionally arrayed Iraqi military changed significantly. The sophisticated coalition weapon systems designed to destroy tanks, armored personnel carriers, artillery, and aircraft were of little use against individuals hiding among the populace, and in fact the use of such systems was often counterproductive. The result was decreased freedom of action for coalition forces, which in turn should have resulted in the coalition's abandoning the direct approach and adopting an indirect strategy much sooner. Unfortunately, this did not happen. For too long the coalition maintained the same operating philosophy it started with and thus contributed to resentment, resistance, and revenge in addition to the view of the coalition forces as occupiers, not liberators.

Conclusion: military strategy failing to serve policy

Wars almost always end long before an opponent's capacity to resist is lost. They end when one side is convinced that more is to be gained through peace than through continuing to fight. Even the stronger of two belligerents may opt for peace when the cost of victory becomes too great. Creating the appropriate psychological state is the essence of strategy. Developing psychological cohesion on one side while disrupting it on the other side is fundamental.

Beaufre's model provides a useful framework for assessing what appears to be US strategy in Iraq. The United States is predisposed to seeking and adopting quick and decisive military solutions, but its military excellence and material advantage almost guarantee that its enemies will resort to irregular means, thereby mitigating an absolute advantage. The American military experience, to a great extent shaped by the ability to marshal manpower, equipment, and industrial capacity, defaults to warfare that is direct in nature regardless of the disposition of the enemy or the international setting. The new American way of war, codified by shock and awe and NCW, had devastating effects on targets but contributed little toward achieving the desired political outcomes. In fact, NCW's perceived benefit of limiting the footprint in theater ceded the stage to our enemies, making the achievement of political outcomes much more costly and perhaps rendering them unachievable. During OIF, the coalition never adequately controlled the countryside. Furthermore, messages are sent by actions as well as by words, and the nature of the "direct" American way of war, which essentially became a manhunt, could not help but alienate the general population. Additionally, force protection measures, though important, separated coalition forces from the general population (up until recently), exacerbating this alienation. The result is a fertile recruiting environment for the insurgents.

Perhaps the most serious error committed by military planners was disassociating political goals from military strategy. The initial freedom of action gained by the Anglo-American coalition resulted in the forging of a hostile anti-US coalition that included many traditional allies. This had at least two effects. First, the burden of the war and its aftermath, in blood and treasure, would rest on the shoulders of the United States. Those countries whose concerns were disregarded at the "takeoff" of the war would not be there at the "landing." Second, dissension within the ranks of the international community regarding the war emboldened the insurgency and gave it a possible wedge against the coalition. Placing military expediency over policy requirements resulted in undermining the occupation.

Finally, the overall cost and duration of the occupation may have moved beyond levels acceptable to the US population. As President Nixon remarked, referring to the war in Vietnam, "When a President sends American troops to war, a hidden timer starts to run. He has a finite period of time to win the war before people grow weary of it."[53]

The clock is ticking. Errors made early on might have been overcome if the US military was better at making strategic adjustments. It took almost a year after the start of OIF before anyone in a significant office acknowledged that the coalition was facing an insurgency. The failure to understand the nature of the threat and, more importantly, the limitations it placed on the direct nature of American strategy led to unnecessary losses and reduced the legitimacy of the new government in Iraq. Rather than adopting an indirect approach, the US military continued its traditional direct approach. Why? Because it could, and the strategic DNA of the US military demands it. But what happens in a situation when only an indirect approach is available?

The Philippines: Mounting sustainable operations against terrorism

The US military was training the Philippine armed forces to more effectively combat terrorism before the terrorist attacks in the United States on 11 September 2001. In Manila and Washington, the events of 9/11 only increased the awareness of the threat and the need to intensify the effort.

The southern Philippines is typical of areas that are ripe for insurgency. It is located along ethnic, cultural, and religious fault lines in a region that has been loosely controlled or governed throughout its long history of occupation. It is home to a discontented Muslim population that has been dominated by a predominately Catholic government based in Manila. The Philippine Muslim population of approximately five million resides primarily in Mindanao and the Sulu Archipelago, some of the poorest provinces of the Philippines.[54]

The area is infamous for civil unrest, lawlessness, terrorist activity, and Muslim separatist movements. It is also home to several al-Qaeda-linked insurgent and terrorist organizations, including the Moro Islamic Liberation Front and the Abu Sayyaf Group (ASG), and it has also been a safe haven for the Indonesia-based Jemaah Islamiyya. The core leadership of many of these groups received their initial training in the camps of Afghanistan and gained experience in the jihad against the Soviet invasion during the 1980s. In fact, Osama bin Laden's brother-in-law provided the ASG with its initial funding. While al-Qaeda is not the cause of these movements, it has used them as a vehicle to expand its global reach and spread its extremist ideology.[55] The United States became interested in the southern Philippines in May 2001, when the ASG kidnapped several US citizens and held them hostage on their island stronghold of Basilan. After 9/11, the region became a key piece in the Global War on Terror when Washington and Manila set their sights on the destruction of the ASG.

The indirect approach

Unilateral operations could not be conducted within the borders of the Philippines, an allied state, without its consent. In fact, the deployment of US forces inside the Philippines was a controversial issue there, and many believed incorrectly that Philippine law prohibited the employment of foreign troops on Philippine soil. Facts are often rendered irrelevant when an issue becomes a political one. Politicians, exploiting the Philippine press, made the employment of US troops so contentious that Washington severely restricted the deployment of US forces and prohibited troops from initiating combat operations. In fact, US personnel could use deadly force only to act in self-defense and in the defense of others. US freedom of action would be severely restricted.[56]

In the whirlwind of activity following 9/11, the US Pacific Command's (USPACOM) Special Operations Headquarters (SOCPAC) developed a training package designed to improve the ability of the Armed Forces of the Philippines (AFP) to defend the country against terrorism. This plan was based on assessments of AFP capabilities conducted by US personnel with considerable experience in the Philippines. The Philippine government affirmed its commitment to combat terror when President Arroyo visited Washington in November 2001 and met with President Bush. The two leaders declared their intentions to combat terror and protect Philippine sovereignty. To this end, the US–Philippine agreement included a $100 million military assistance package, $4.6 billion in economic aid, and the deployment to the Philippines of US advisory teams. The Bush–Arroyo meeting established the policy objective of defeating the terrorists in the Philippines by enhancing local capabilities.[57]

Another nontrivial aspect of the US effort was that the Philippines remains an "economy of force" theater. With the US focus concentrated in Iraq and Afghanistan, US military forces were already largely committed, leaving very few available for operations in the Philippines. There were other shortfalls. The priority for combat-related equipment, supplies, and ammunition was to Iraq and Afghanistan, again leaving little to support operations in the Philippines. The relatively low resource priority given to the Philippines was compounded by the higher priority given to China and North Korea by USPACOM. Viewing the situation in terms of Beaufre's model is instructive:[58]

- Freedom of action was very limited. US restrictions as a result of Philippine politics severely reduced the engagement options for US forces.
- The importance of the issue was moderate. Although the Global War on Terror was and remains a priority, actions in the USPACOM theater reveal a clear geographic focus on the Middle East.
- Resource availability was limited. Priority operations in Iraq and Afghanistan consumed almost all available forces, equipment, and supplies.

Logic, along with a clear absence of any reasonable alternative, dictated an indirect approach. USPACOM tasked SOCPAC to plan and execute the "Global War on Terror Southeast Asian Campaign." Based on an analysis of the situation, SOCPAC determined that the best way to proceed was to focus on three interconnected activities:

1 Capacity building for the Armed Forces of the Philippines: US military forces would train, advise, and assist Philippine security forces to better create and maintain a secure and stable environment.
2 Civil–military operations: Humanitarian and civic action projects led by the

Philippine government and facilitated by the United States would improve quality of life in the affected area and demonstrate the Philippine government's concern for its citizens.

3   Information operations: The successes of the above two activities would be exploited in the minds of the people to enhance the legitimacy of the government.[59]

The underlying principle in this strategy is that, for the people living in the affected area, the expected value of supporting the Philippine government must be made to exceed the expected value of supporting the terrorists.[60]

In February 2002, the United States dispatched a contingent of US troops to the Philippines. The tip of the spear was 160 Special Forces deployed into terrorist-controlled Basilan. This force was later augmented with Navy and Marine construction units to assist with civic action projects on the island. The Special Forces advisers lived and worked with their AFP counterparts, who were scattered around the island in remote areas located near ASG strongholds. The AFP units on Basilan were unorganized, lacked training in basic infantry skills, and did not aggressively pursue the ASG.[61]

The Special Forces teams developed close professional relationships based on mutual trust with their AFP counterparts. Equally important, they also developed rapport with the local villagers, using their language and cultural skills to interact with them in routine activities. Special Forces medics played a key role in establishing initial trust among the population by providing medical assistance alongside the AFP in remote villages. Special Forces advisers also met with key leaders, both military and civilian, on their ground, to explain the rationale behind the Philippine–US effort, to dispel rumors, and to manage expectations. Simultaneously, the Special Forces teams were increasing their own situational awareness in order to facilitate tactical adjustments, measure effects, and enhance force protection. Presence, perseverance, and patience characterized the operating philosophy.[62]

The most important task was to protect the local population by establishing a secure environment. This required that the APF be retrained to increase unit proficiency and instill confidence. The Special Forces teams immediately went to work training Philippine soldiers and marines in basic combat skills. The Philippine troops were also trained in giving emergency medical treatment in the field, which proved to be a significant morale booster. At the same time, the poorly defended AFP base camps were converted into tactically defensible areas. As unit proficiency improved, security patrolling increased; increased patrolling by competent units led to seizing the initiative from the insurgents and establishing security at the village level. Special Forces teams played a key role in building AFP capacity by accompanying units as advisers on combat operations.[63] The establishment of security and the protection of the local population provided the foundation for all other activities designed to establish stability.[64]

Humanitarian assistance and civic action projects designed to meet the basic needs of the local population began once security was established. As the security situation improved, customized projects tailored for particular regions and provinces followed. The Philippine government was in the lead, with the United States providing support. When possible, locally procured materials and workers were used on projects to put money directly into the local economy. These efforts enhanced the legitimacy of both the government in Manila and the AFP and helped restore law and order to the island. Most importantly, the projects reduced support for the terrorists in Muslim villages on Basilan, enabling the AFP to cultivate closer relations and strengthen their control over the local population in insurgent-influenced areas.[65]

Stability and security on Basilan Island vastly improved from the days when Abu Sayyaf gunmen staged kidnappings and killings that held the entire island hostage. According to *Mindanews*, "Every day, someone was killed. The streets of Isabela became a battleground."[66] In 2001 alone, a doctor working on the island recorded eighty-two kidnappings, thirty-two deaths by gunshot wounds, and nineteen beheadings.[67] In 2003, the same doctor recorded only five autopsies and not a single kidnapping or beheading. Insurgent and extremist activity dropped dramatically. Today, the majority of citizens on the island, Muslim and Christian alike, no longer live in fear. Residents of Isabela, Basilan's capital city, stroll the Plaza Rizal in the evening to hear live music. Government officials say that the islanders now readily provide information to thwart insurgent attacks. Businessmen have begun to invest on the island. Finally, improved security conditions on Basilan have resulted in the severing of islanders' ties with extremists.[68]

Eliminating the ASG as an imminent threat has resulted in a 70 percent reduction in military forces deployed on Basilan. Prior to US involvement in 2002, the Philippine government had seven infantry battalions deployed on Basilan, battling Abu Sayyaf and struggling to maintain law and order. Since 2003, the AFP has needed just two US-trained infantry battalions on the island and has transitioned many security-related responsibilities to local police and militia units. All this occurred as a result of improved operational capability on the part of AFP and an enhanced role on the part of the Philippine government in improving the quality of life of the people of Basilan. These activities have permitted the AFP to assume a more prominent role in peacekeeping and development, in contrast with a past that emphasized force-on-force combat operations against the terrorists or insurgents. In short, Basilan Island is very different today than prior to 2002.[69] The physical landscape of the island remains largely unchanged. The rugged terrain and remote villages that extremists once found so inviting and conducive to their deadly activities are all still there. What has changed is the attitude and loyalties of the Basilan people, making the environment far less conducive to insurgent or terrorist activity.

Although the ASG was not destroyed, the Philippine–US effort produced notable results, illustrating a positive relationship between policy, strategy, and outcomes. The results included denying insurgent and terrorist sanctuary on Basilan Island; improving the capacity of the AFP; enhancing the legitimacy of the Philippine government in the region; establishing the conditions for peace and development on the island; and providing a favorable impression of US military efforts in the region.[70]

Conclusion: military strategy serving policy

The policy objective that emerged from the November 2001 Bush–Arroyo meeting was clear. The Global War on Terror efforts in the Philippines required the United States to assist the Philippine government in defeating the terrorists by developing and enhancing local capabilities. The United States was restricted in how US forces would prosecute this portion of the Global War on Terror in at least two ways. First, as a sovereign nation and ally, the Philippine government limited the scope of US involvement. Philippine politics would not permit the US presence to overwhelm Philippine actions. Sensitive to Philippine domestic concerns, the United States also restricted US involvement in going after terrorists in the Philippines. Second, the priority efforts in the Global War on Terror were in Iraq and Afghanistan, which meant that despite the overall importance of the global anti-terror effort, the Philippines was an "economy of force" theater. Manpower, equipment, supplies, and high-level attention were visibly absent. Counterintuitively, this opened the door to success.

There were significant advantages to the imposed restrictions and not being in the spotlight. The American way of war, characterized historically by the ability to marshal manpower, equipment, and industrial capacity, and more recently by network-centric warfare and shock and awe, did not have an opportunity to come into play. The absence of these key features that the United States counts on to wage war necessitated a different approach. The default strategy that relies on sophisticated technology and material superiority, regardless of the nature of the threat, was simply not an option. Fortuitously for the United States, SOCPAC clearly understood the operational environment, the need to engage the threat discriminately and indirectly, and the imperative of ensuring the legitimacy and credibility of the actions. This keen understanding resulted in a plan that anticipated and controlled the psychological effects of war.[71]

Creating the appropriate psychological state is the essence of strategy. The results of the Philippine–US effort compares nicely with the list of things to avoid presented earlier in the analysis of Iraq – losing control of the countryside, alienating the general population, and so on. The successful strategy that was executed in the Philippines is consistent with the patterns of strategy outlined by Beaufre. The conditions required an indirect response; however, the requirement for an indirect response is no guarantee of getting one. One cannot overemphasize the happy coincidence of having a "right-minded" chain of command in SOCPAC and a disengaged Pentagon and theater command. It is interesting to contemplate what the situation in the Philippines would be like if US military efforts had not been focused on Iraq and Afghanistan and if the Philippine government had been willing to accept a larger direct role by American forces. To start with, the right-minded SOCPAC chain of command would have been displaced. The renewed freedom of action resulting from manpower and material superiority coupled with the Philippine government acceptance of a larger and more direct US role would likely have generated a strategy similar to those that unfolded in Iraq and Afghanistan, in spite of natural constraints on freedom of action based on geography, politics, culture, history, ethnicity, and religion. It is likely that Basilan Island would have become a magnet for jihadists.

Conclusion and recommendations

Strategy is the fabric that binds the broad range of US political, military, economic, and psychological actions.[72] Properly conceived and executed, military strategy facilitates the effective and efficient attainment of policy objectives by reducing the morale and combat efficiency of enemy troops and creating dissidence and disaffection within their ranks. Good strategy also promotes resistance within the civilian population against the enemy while simultaneously increasing the legitimacy of one's own cause. It is not simply the objective of good PSYOP to achieve an end state where enemy, neutral, and friendly actors will take action favorable to the United States and its allies; it is the ultimate objective of good strategy to create these conditions. Even the best PSYOP cannot rescue ill-conceived strategy.

Strategy includes the integration of all the instrumentalities available to a state. The notion of subordinate, independent strategies, or substrategies, undermines its essence. Sound strategy is necessary to increase the probability of achieving desired outcomes. But even sound strategy is no guarantee of success against foolish policy or military incompetence.[73] Those responsible for the security of the United States should see it as their duty to find ways of avoiding war, and failing that, to make war cost as little as possible. Strategy links means with ends. It is a process through which a nation attempts to minimize its weaknesses and limitations and maximize its strengths and capabilities in the international arena. All states possess four traditional instruments to pursue policy objectives – diplomatic, economic, military, and psychological.[74] Although only rarely must all four be brought to bear in a particular situation, only at great risk can a nation overlook or fail to use each of them. What scholars and theoreticians call "total strategy" or "grand strategy" encompasses an understanding and willingness to use all of these instruments in a coordinated fashion.

The first three instruments are generally recognized, but there is still argument about the equality of the fourth. Military and economic instruments can be described as physical in nature; we can measure and set relatively precise limits on their application. Diplomatic and psychological instruments, by contrast, cannot be measured, at least not with precision, since they are often neither seen nor felt. Indeed, they fall in the realm of the intellectual and the emotional, which makes analysis difficult – and makes civilian and military leaders uncomfortable.[75]

There is another analytical distinction that stems directly from the separation of these instruments into the physical and the intellectual-emotional. Since PSYOP is carried out in the minds and hearts of people, it transcends military and economic interactions. The term nation is meaningless unless it refers to the human element that makes it up. Nations do not oppose policies of other nations, people do. And the moment we accept that people are the actors and reactors in international affairs, we must accept the fact that these people are guided by their minds and emotions. If this is true, the manipulation and countermanipulation of the physical instruments are subordinate to the intellectual-emotional.[76]

With this in mind, I conclude this chapter with two recommendations. First, the mindset that psychological warfare is the province of specialists and is somehow separated from "real" strategy must change. This does not mean that a requirement for specialists does not exist, but PSYOP must be an integral part of strategy. Second, a system is needed to identify people, civilian and military, who are inclined to think strategically, along with a method of sharpening the minds of these individuals. I am not sanguine about the ability of the United States to recognize these shortfalls, let alone to take corrective steps. Fortunately, at least for the time being, we can overcome strategic blunders without serious adverse effects on our quality of life. There is no assurance that this condition will persist.

Notes

1 *Review of Psychological Operations Lessons Learned from Recent Operational Experience* (Washington: National Defense University Institute for National Strategic Studies, 2005), 2; hereafter cited as *Lessons Learned.*
2 See *Lessons Learned*, 45; Charles H. Briscoe, Richard L. Kiper, James A. Schroeder, and Kalev I. Sepp, *Weapon of Choice: ARSOF in Afghanistan* (Fort Leavenworth, KS: Combat Studies Institute Press), 113.
3 See *Lessons Learned*, 46; United States Central Command, "Campaign Plan for *Enduring Freedom*: Information Operations," appendix 3 to annex C (Operations), C-3–8.
4 *Lessons Learned*, 46.
5 Ibid., 47.
6 Ibid.
7 See Briscoe *et al.*, *Weapon of Choice*, 47.
8 *Lessons Learned*, 48.
9 Ibid.
10 Ibid., 48, 51.
11 Ibid., 49.
12 Ibid., 50.
13 Ibid.
14 Center for Army Lessons Learned (CALL), "On Point: The United States Army in *Iraqi Freedom*" (Fort Leavenworth, 11 August 2004), chapter 4.
15 *Lessons Learned*, 50.
16 See *Lessons Learned*, 53–5; "Allied Views on PSYOP Operational Lessons Learned," commentary from allied information operations experts in response to inquiry on US PSYOP performance in Operation Enduring Freedom and Operation Iraqi Freedom, 27 August 2004: "On occasions US PSYOP appeared to be too focused on 'warfighting' issues rather than wider issues and could therefore be too direct and aggressive to the detriment of the overall long-term mission objectives."
17 See cable from USCENTCOM MACDILL AFB to RUFDAVC/CDRVCORPS HEIDELBERG GE//J3/IO//RUESOC/COMJPOTF CC FT BRAGG NC; Subject: CFC FRAGO 09–304 to OPORD 09; *Lessons Learned*, 55.
18 *Lessons Learned*, 56.
19 Ibid.

20 See Center for Army Lessons Learned, "Operation *Iraqi Freedom* (OIF) CAAT II Initial Impressions Report (IIR)," Report No. 04–13 (May 2004), 4, 57.
21 Ibid., iii, 5, 58.
22 *Lessons Learned*, 14–15.
23 Ibid., 16, 19.
24 Ibid., 19–20.
25 The Department of Defense Directive on Information Operations defines information operations as "The integrated employment of the core capabilities of electronic warfare, computer network operations, psychological operations, military deception, and operations security, in concert with specified supporting and related capabilities, to influence, disrupt, corrupt, or usurp adversarial human and automated decision-making while protecting our own."
26 Office of Strategic Influence, en.wikipedia.org/wiki/Office_of_Strategic_Influence; Defense Department news briefing, Secretary of Defense Donald H. Rumsfeld, 26 February 2002.
27 See Allen Pusey, "Propaganda Part of War, Experts Say," *Monterey Herald*, 27 December 2005, A1; and Reuel Marc Gerecht, "Hearts and Minds in Iraq," *Washington Post*, 10 January 2006, 15.
28 Gerecht, "Hearts and Minds in Iraq."
29 Ibid.
30 André Beaufre, *An Introduction to Strategy* (New York: Praeger, 1965).
31 George E. Orr, *Combat C3I Fundamentals and Interactions* (Alabama, Air University, Maxwell Air Force Base, 1983), Research Report No. AU-ARI-82-5.
32 Beaufre, 34–6.
33 Ibid., 57.
34 Ibid., 22.
35 Ibid., 110.
36 Ibid., 57–8.
37 Beaufre actually discusses five patterns of strategy. I have condensed them into four because the first and fifth both use direct military action based on having sufficient resources available.
38 Beaufre, 26–7.
39 Ibid., 28.
40 Ibid., 29–30.
41 Frederick W. Kagan, "War and Aftermath," *Policy Review*, Aug–Sep 2003, 1.
42 Thomas Donnelly, *Operation Iraqi Freedom: A Strategic Assessment* (Washington, DC: American Enterprise Institute Press, 2004), 29.
43 See Kagan, 2–3. Also, the best discussion on NCW is given by David Alberts, John Gartska, and Frederick Stein in a book called *Network Centric Warfare: Developing and Leveraging Information Superiority* (CCRP Publications, second edition, 1999).
44 Kagan, "War and Aftermath," 3–7.
45 For a thorough discussion of the limits of airpower see Robert A. Pate, *Bombing to Win: Air Power and Coercion in War* (Ithaca, NY: Cornell University Press, 1996).
46 See George Packer, *The Assassins' Gate: America in Iraq* (New York: Farrar, Straus and Giroux, 2005), 137; Kagan, "War and Aftermath," 6.
47 Packer, *Assassins' Gate*, 136.
48 Hy S. Rothstein, *Afghanistan and the Troubled Future of Unconventional Warfare* (Annapolis, MD: Naval Institute Press, 2006), 170.
49 Jeffrey Record, *Dark Victory: America's Second War Against Iraq* (Annapolis, MD, Naval Institute Press, 2004), 64–9.
50 Ibid., 91–3.
51 Ibid., 68–9.
52 Ibid., 93.
53 Richard Nixon, *No More Vietnams* (New York: Arbor House, 1985), 88.

54 Greg R. Wilson, *Operation Enduring Freedom – Philippines: "The Indirect Approach"* (USAWC Civilian Research Project, 2006), 5–6.

55 See Wilson, *Operation Enduring Freedom – Philippines*, 6; David S. Maxwell, "Operation Enduring Freedom – Philippines: What Would Sun Tzu Say?" *Military Review*, May–June 2004, 22.

56 Cherilyn Walley, "Impact of Semipermissive Environment on Force – Protection in Philippine Engagements," *Special Warfare*, Summer 2004, 36.

57 C. H. Briscoe, "Balikatan Exercise Spearheaded ARSOF Operations in the Philippines," *Special Warfare*, September 2004, 17–8.

58 Ibid., 18–19; interviews conducted by the author while in the Philippines and USPACOM Headquarters, Hawaii, November 2005.

59 See Wilson, *Operation Enduring Freedom – Philippines*, 10; Briscoe, "Balikatan Exercise," 18; Col. Linder, interview by the authors, November 2005, JSOTF-PI, Manila, Philippines.

60 Gordon H. McCormick developed a counterinsurgency model that accounts for the domestic and international environments, including the dynamic relationships among the government, insurgents, people, and international actors. Variations of this model have been taught to Special Operations Officers at the Naval Postgraduate School for many years. Fortunately, a few of these officers were present at SOCPAC to influence the plan. See Gordon H. McCormick, "A 'Pocket Model' of Internal War," Department of Defense Analysis, Naval Postgraduate School, forthcoming.

61 Ibid., 12.

62 Ibid., 13.

63 Special Forces advisers were initially not allowed to accompany AFP units below the battalion level. This eventually changed and Special Forces advisers were allowed down to the company level.

64 Ibid.

65 See Cherilyn Walley, "Civil Affairs: A Weapon of Peace on Basilan Island," *Special Warfare*, September 2004, 31–4; Wilson, *Operation Enduring Freedom – Philippines*, 13–14.

66 Froilan Gallardo, "Balikatan Over But Abu Sayyaf Problem Far from Over," *MindaNews*, 29 December, 1.

67 Thomas Crampton, "US Aids Fragile Peace in the Southern Philippines," *International Herald Tribune*, 28 February 2003.

68 Wilson, *Operation Enduring Freedom – Philippines*, 16.

69 Ibid., 17.

70 Ibid., 20.

71 The strategic culture in portions of the Special Operations community is fortunately inconsistent with that of the conventional military. This difference facilitated the development of the successful strategy for the Philippines.

72 I use the term "psychological" in the broadest sense. The terms "informational" or even "ideological" are equally appropriate.

73 Colin S. Gray, *Irregular Enemies and the Essence of Strategy: Can the American Way of War Adapt?* (Carlisle, PA: Stragegic Studies Institute, 2006), 50.

74 These four traditional instruments of statecraft – diplomatic, economic, military, and psychological – are referred to today as elements of national power and now include the following instruments: military, informational, diplomatic, law enforcement, intelligence, finance, and economics. It is debatable whether the newer formulation is an improvement over the traditional four instruments.

75 R. W. Van de Velde, "Instruments of Statecraft" (Washington, DC: US Government Printing Office, ST 33–151, 1963), IIA2–IIA5.

76 Ibid.

8 Assessing the computer network operations threat of foreign countries

Dorothy E. Denning

As the introduction to this book so aptly stated, advances in information technologies simultaneously empower and imperil those who use them. They empower by facilitating communications and the flow of information; they imperil by introducing new vulnerabilities and targets of attack. Information strategy has to adapt to both of these effects, exploiting and leveraging the enabling technologies while protecting against threats to the very same technologies we come to rely upon.

In this chapter I address the latter – the defensive side of information strategy as it applies to computer and networking technologies. Computer networks have become the target of an ever increasing number of hackers, criminals, spies, and others who have found advantage in exploiting and damaging them. These actors penetrate computer networks in order to steal, degrade, and destroy information and information systems. They launch computer viruses and worms, conduct denial-of-service attacks, vandalize Websites, and extort money from victims. The effects have been costly: businesses disrupted or closed, military systems disabled, emergency and banking services suspended, transportation delayed, military and trade secrets compromised, and identity theft and credit card fraud perpetrated around the globe. The potential consequences of cyber attacks will only get worse as our use of and reliance on information technologies increase.

Many government officials and security experts believe that foreign governments pose the largest threat to computer networks, followed by terrorists. Of especial concern is a possible "electronic Pearl Harbor" or act of cyber terrorism that would affect a critical infrastructure such as the power grid or banking network, with devastating economic, social, or national security consequences. An attack against military networks could potentially undermine the armed forces' ability to effectively fight an adversary, especially during a time of conflict, and even attacks against civilian infrastructures, such as energy and telecommunications, could severely damage military capability because of widespread dependence on civilian systems.

So far, the number of reported cyber attacks attributed to foreign governments or terrorists has been relatively small, and none have been devastating. Cyber incidents attributed to governments have mostly involved espionage, and network attacks by terrorists and their sympathizers have fallen more in the domain of crime and vandalism than terrorism – mainly Web defacements, denial-of-service attacks, and credit card fraud. In 1999 and early 2000, the Chinese government was accused of attacking foreign Websites associated with the outlawed group Falun Gong,[1] but government sabotage of this type against foreign computers appears to be the exception. Today it seems more likely that the Chinese government would use its national firewalls to filter out objectionable Websites than launch attacks against them. However, government exploitation of computer networks for intelligence purposes seems highly likely given intelligence exploitation of other telecommunications media.

The paucity of published information about what terrorists and governments are interested in and able to do in cyberspace, coupled with the fact that nothing resembling an electronic Pearl Harbor or act of cyber terrorism has occurred, has led many to question whether these threats have been overhyped or are even real. Yet, it would be as foolish to dismiss such threats as it would be to base policy and plans on speculation and unsubstantiated fear. Instead, we need well-grounded assessments of what potential adversaries are motivated to do and capable of doing.

We also need sound assessments of vulnerabilities in critical infrastructures and how risks can be mitigated. However, these evaluations can be conducted without regard to any particular actor or motive. Computer networks need to be protected from damaging attacks regardless of whether they originate from a runaway worm, a hacker out to see what's possible, a greedy crook who sees an opportunity for extortion, a former employee seeking revenge, a nation-state, or a terrorist. Computer worms alone have brought down emergency 911 services, a train signaling system, the safety monitoring system at a nuclear power plant, and ATM networks. Insiders determined to cause harm are in a particularly powerful position. In what was perhaps the most damaging infrastructure attack, a former contractor, armed with the requisite hardware, software, and knowledge, hacked a water treatment system in Australia and caused raw sewage overflows.[2]

Arguably, it may be more important to focus on protecting the networks rather than studying particular actors. However, there are also benefits to be gained by understanding the motives and capabilities of those who might attack them. First, if networks are attacked, we would be in a better position to narrow down likely perpetrators. Second, if we enter into military conflict with a particular adversary, we would know what that adversary could and could not do to our military networks and critical infrastructures. Third, we may learn of capabilities and methods of attack that we had not considered.

In 2003, the Naval Postgraduate School began a study to assess the computer network operations (CNO) threat of foreign countries. The objective was to develop a general methodology that could be applied to any country and to apply it to specific countries as test cases. For our study, we chose Iran and North Korea. Our country results were published in two master's theses, one on Iran[3] and one on North Korea.[4]

In our project, we sought to elaborate a comprehensive methodology and were less concerned about producing a thorough, definitive assessment of the countries we chose. Indeed, because we limited our research to unclassified information available through open sources, we almost certainly missed key information about these countries. We did not attempt to determine what the intelligence services might know that we did not.

This chapter summarizes the results of our research. The next three sections describe our methodology and the results for Iran and North Korea. In the country sections, citations are to original sources where verified or found in the process of writing this paper. Otherwise, citations are to the theses. I have also added some of my own thoughts, which are presented without citation.

In the discussion of Iran especially, I have singled out specific individuals and groups who have engaged in CNO-related activities to illustrate the capability we found. In so doing, I do not mean to imply they are the only ones working in CNO or that they pose any sort of threat – indeed, many are working toward better information security.

Methodology

The US Department of Defense defines computer network operations (CNO) as comprising three types of operations: computer network attack, computer network defense, and related computer network exploitation-enabling operations.[5] Computer network attack (CNA) refers to operations to disrupt, deny, degrade, or destroy information resident in computers and computer networks, or the computers and networks themselves. Computer network exploitation (CNE) consists of enabling operations and intelligence collection to gather data from target or adversary computers and networks in support of CNA. Computer network defense (CND) consists of defensive measures to protect and defend information, computers, and networks from disruption, denial, degradation, or destruction. In short, CND refers to operations that protect against adversary CNA/E.

Outside the US military, it is common to use the term "attack" to refer to any operation that intentionally violates security policies and laws. This includes CNE as well as exploit operations conducted for the purpose of intelligence collection, not just to enable CNA. It is also common to see the term "security" for "defense" and to include within it protection against adversary intelligence operations as well as information operations that disrupt, deny, degrade, or destroy. Both sets of terms are used in the discussion below.

Although the CNO threat is derived from attack/exploit operations rather than defense, we included the latter in our analysis. It is not possible to build strong defenses without knowledge of how systems are attacked, so the presence of a CND capability within a country implies at least some CNA/E knowledge. Furthermore, it seems unlikely that any country would develop a CNA/E capability if it is unable to defend its own networks from a counterattack, so the apparent lack of a CND capability would suggest a corresponding lack of CNA/E capability, assuming no information to the contrary. A country with a strong CND capability would be in a much better position to build and use a CNA/E capability than one without.

To assess a foreign state's CNO threat, we looked for indicators of capability and intent to conduct CNO. These indicators were based on generic factors that could be applied to any country. The factors were grouped into four general categories:

- Information technology industry and infrastructure.
- Academic and research community.
- Government and foreign relations.
- Hacking and cyber attacks.

The categories are not entirely disjointed. For example, government-sponsored research on CNO falls into the second and third categories, and government-sponsored cyber attacks fall into the third and fourth. In the discussion below, we have generally assigned each type of activity to a single category and treated it in that context.

Within each category, we began with an initial set of questions to guide our search for information, although we did not limit our collection to those questions. In many cases, we could not answer the questions directly but found other information that was useful for our analysis.

All of the information we used was unclassified. Most was acquired through the Internet and personal contacts in the United States and countries other than those we studied. A more complete picture could be obtained with access to classified information or to persons within the countries of study. It was especially difficult to obtain information that originated in North Korea owing to the closed nature of the country and its apparent isolation from the Internet.

Most of the material we used was in English. We arranged for translation of a few Web pages in Farsi that we thought might be useful for the Iranian study, but limited time and resources precluded translating more. For the most part, we simply ignored Websites and documents that were not in English. A comprehensive study that includes more foreign language sources could very well turn up evidence we did not find.

We made extensive use of Google searches to find relevant information. These searches led us to Web pages that we had not found by simply browsing institutional Websites. However, we did not have time to pursue all search hits or to try an extensive set of search strings, which leaves open the possibility that we missed a large amount of useful information. For future study, it would be worthwhile to try to identify a collection of search strings that would likely uncover most of the relevant information that can be found through open-source searches.

We began our study in October 2003. Two students were assigned to the project, one for Iran and one for North Korea. Their objective was to report their results in the form of separate master's theses for a September 2004 graduation.

In March 2004, we were invited by the Institute for Security Technology Studies at Dartmouth College to review a draft report of a study whose objectives were very close to our own. At that time, we were far enough along with our own work to provide feedback on theirs, and their study provided valuable input for our own. Their final report was published in November 2004,[6] after our study of North Korea was complete but while we were in the middle of our Iranian study. The Iranian study was delayed for a year because the student conducting the work had an unexpected reassignment of duties. Another student later joined the project to see it through to a September 2005 completion.

The Dartmouth group took a somewhat different approach in their analysis. In particular, they organized evidence indicative of capability or intent into two categories. Category 1 evidence consists of direct links to a foreign cyber warfare capability. It is derived from US government reports (which we did not use), foreign official statements, and foreign military and intelligence agency research. Category 2 evidence consists of circumstantial links indicating a baseline information technology infrastructure necessary to support a cyber warfare operation. The Dartmouth country reports are organized around these categories and sources of evidence, whereas ours are organized around the four categories of activity described above. The Dartmouth report also covers six countries, including the two we studied. Besides Iran and North Korea, they studied China, India, Pakistan, and Russia.

The following subsections describe the four areas of activity we investigated, our rationale for looking at these areas, and the type of information we sought.

Information technology industry and infrastructure

Our goal here was to assess a country's information technology (IT) industry and its information infrastructure. In the area of IT, we examined the country's hardware and software industry, IT service companies, access to international IT supply chains, industry partnerships with foreign companies, and IT professionals in the country; we paid particular attention to companies that provided CNO-related technologies or services. However, other areas of technology are also relevant. If a country does not have access to or experience using popular hardware and software platforms, such as Microsoft Windows and related products, it will be at a disadvantage in terms of developing a capability to attack or defend those systems. Also, many of the skills used in one area of IT, such as general knowledge and skills in computer networks, operating systems, and programming, are transferable to CNO.

In the area of infrastructure, our main interest was computer networks, especially the Internet and intranets, but we also considered the country's telecommunications and electrical infrastructures, since both support networking. If telecommunications or electricity is inadequate or unreliable, it may be difficult to launch a sustained attack against another country.

For computer networks, we considered prevalence, connectivity, capacity, technologies and platforms used, presence of Internet service providers (ISPs), and government regulations. We reasoned that a country that is well connected through modern technologies and high-speed links is in a better position to develop a CNO capability than one that is not, as it can draw on the considerable expertise and talent acquired through use of the networks. Internet penetration is particularly valuable, because it gives the population access to global CNO resources, such as hacking tools and "how to" guides, as well as to international targets to attack. But even a country that has promoted a national intranet while stifling or prohibiting Internet use is in a better position than one that has little networking of any type.

We also considered the legal and regulatory infrastructure as it pertains to CNO, including computer crime laws and their enforcement. A lack of laws in this area could be indicative of little hacking activity within the country or against the country's computer systems, in which case one might conclude that the country has little or no CNO capability, offensive or defensive, at least outside government. However, an absence of cyber crime laws might also mean that more general laws (e.g. governing sabotage and fraud) are considered sufficient for prosecuting cyber attacks.

Academic and research community

For this category, we assessed the extent to which faculty and students engaged in educational and research activities that support a CNO capability. We also examined research conducted in the public domain by persons outside the academic community. Research areas we examined include system and application vulnerabilities, computer crime and network attacks, technologies and methods of defense, and CNO policy and legal issues. Within this broad research community, we looked for CNO-related publications and projects. We also looked for conferences and workshops hosted by members of the community within the country or attended by members of the community in other countries.

In the academic community, we focused on higher education. We looked for courses in areas of CNO and for faculty and students who were conducting research or publishing papers related to CNO. We tried to determine whether any faculty members who were engaged in CNO activity had been educated outside their country and whether students studied CNO abroad. Much of our information was obtained by searching online for school Websites and résumés containing CNO-related entries.

We also examined general education in IT and the IT skills of students at all levels, including primary and secondary school. We were especially interested in whether college students in the country participated in the annual ACM International Collegiate Programming Contest,[7] and if so, how well they did. The ACM programming contest, which traces it roots to a competition at Texas A&M University in 1970, has evolved into a multitiered competition involving three-person student teams from around the world. In 2005, the contest drew 4,109 teams from 1,582 universities in seventy-one countries. We reasoned that a country needs talented programmers to develop new or sophisticated cyber attacks, so placing well in the contest would suggest the presence of a talent base on which to draw.

Government and foreign relations

Here we considered efforts on the part of government agencies to develop a CNO capability. We looked for signs that the government was creating one or more CNO units or teams, conducting training in CNO, or sponsoring or conducting research on CNO. We also looked for documents or statements from official government sources that outlined government policy or doctrine on CNO.

We tried to determine whether the government was using the Internet for intelligence collection, and if so, whether its tactics went beyond open source collection to hacking into computer networks. We reasoned that a government with the ability to penetrate and exploit foreign networks for intelligence collection would have a head start on developing a CNO capability, as many of the same skills are needed.

We considered a government's relations with other countries to determine whether it might acquire CNO-related resources from another country. Such resources might include information, technology, or training in CNO. We also looked for motives and objectives that might lead the government to conduct a cyber attack against another country.

Hacking and cyber attacks

This category focused on actual cyber attacks originating in the study country. We considered attacks by all types of actors, from teenage hackers to criminal groups to government agencies. (Often, however, it is not possible to determine the source of an attack.) We tried to identify nongovernment hacking groups and individual hackers operating from within the country.

We wanted to know what types of attack the country's hackers conducted and what tools and methods they used. We considered all types of attack, including denial-of-service attacks, Web defacements, launching of viruses and worms, use of Trojan horses and spyware, and so forth. We considered cyber operations that acquired sensitive information, including trade secrets, personal information such as Social Security or credit card information, and sensitive government information.

Although we examined attacks against international targets, we were especially interested in cyber attacks against US systems and whether such attacks were politically motivated. Patriotic Chinese hackers, for example, attacked US systems after the 1999 US bombing of the Chinese embassy in Belgrade during the Kosovo conflict and then again in the wake of the US–China spy plane incident in 2001.

We tried to determine how the government responded to hacking by its citizens. We wanted to know if specific attacks, particularly those against US systems, were tolerated, encouraged, or even supported. We wanted to know if the government hired hackers or otherwise made use of hackers' expertise or skills. A country with an active hacking community can draw on that community to develop its CNO capability; it could recruit them into the military or an agency with CNO authority, employ their services as consultants or trainers, or participate in their activities, such as conferences and online discussion groups. However, if hackers are hired, there is a risk that they will attack the government's own systems or otherwise engage in illegal or inappropriate hacking. A country might also encourage its hackers to participate in a war against another state as "citizen cyber soldiers."

Iran

Jason Patterson and Matthew Smith, both lieutenants in the US Navy, performed our study of Iran. They found considerable amounts of information on Iranian Websites, particularly sites associated with universities, government-sponsored research centers, hacking groups, and industry. Although they concentrated their efforts on sites that were in English, they obtained translations of a few sites that were in Farsi. They completed their study in September 2005.[8] The following subsections summarize some of their key findings and provide additional information and analysis not included in their thesis.

IT industry and infrastructure

Iran's information infrastructure has been undergoing growth and modernization since the first of a series of five-year plans adopted by Parliament in 1990. The plan, which aimed to restore the Iranian economy in the wake of the Iran–Iraq war, included requirements for information and communications technology.[9]

As of 2003, Iran had about twenty-seven main telephone lines and cellular subscribers per 100 people, which represented a 670 percent increase from 1990, when there were only four such lines and cellular subscribers. However, the numbers are still low compared with, say, the United States, which had 117 per 100 population in 2003.[10]

Iran provides access to the global telecommunications network through fiber-optic and satellite links. A 721 km segment of the Trans-Asia-Europe Project, the world's largest overland fiber-optic system, passes through Iran, transmitting data at 622 megabytes per second. In addition, an underwater link transmitting at 140 megabits per second connects Iran to the United Arab Emirates. Satellite communications were achieved with Inmarsat land earth stations connected to commercial satellites, although Iran is now in the process of creating its own satellite network, to include two Russian-supplied Zohreh satellites, five land stations, 135 primary and secondary stations, twenty-seven zonal stations, thirty-one community stations, and 1,374 rural stations.[11]

Iran's foray into the Internet began in the early 1990s when the Institute for Studies in Theoretical Physics and Mathematics joined BITNET through Iran's membership in the Trans-European Research and Education Networking Association. As BITNET was absorbed into the Internet, the Iranian node developed into a Class C Internet node. By 2000, Iran had over thirty ISPs.[12]

According to the International Telecommunications Union, the proportion of Internet users in Iran rose from 1.6 percent in 2001 to 7.2 percent in 2003.[13] By December 2005, it was up to 10.8 percent, or about 7.5 million people.[14]

While promoting the Internet, the Iranian government also censors it. This is done largely under the wide-ranging Press Law of 1986. According to a study by the OpenNet Initiative, the government blocks access to most pornographic sites and anonymizer tools, a large number of sites with gay and lesbian content, some politically sensitive sites, women's rights sites, and certain targeted Weblogs (blogs).[15] The study did not examine whether any hacking sites were blocked. ISPs use filtering software developed in the United States to block foreign sites. Sites based in Iran may be shut down, suspended, or filtered. Operators and authors are subject to pressure and even arrest.

Iran's hardware and software industries are wanting, hampered by state controls, restrictive trade policies, external trade embargoes, contradictory legislation, and a lack of software management expertise within the industries themselves. Iran has approximately 200 companies involved in software development and 20,000 workers in the software industry.[16]

In the area of CNO, we identified one company, Sharif Secure Ware, that bills itself as a network security and consultation company.[17] We also found a software development company, Systems Group, that formed an alliance with a German security company, Securepoint Security Solution. Under an arrangement announced in July 2005, Systems Group will be the exclusive representative of Securepoint products and services in Iran. Together the two companies seek to become the leading Iranian security software company. With almost 600 employees and 4,500 customers, Systems Group claims to be the largest software corporation in Iran.[18] That Systems Group would team with a German security company suggests that Iran might not have a competitive domestic security company, although there could be other reasons behind the partnership.

Iran does not have any laws that define or specifically prohibit cyber crimes. There are copyright protections for domestically produced software, but the laws are seldom enforced and do not apply to imports. Software pirating and hacking both run rampant.[19]

Academic and research community

Iranian universities have strong IT programs, including computer science and computer engineering. They have been active in the ACM programming contest, and two universities did as well as any US schools in the world finals held in Shanghai in April 2005. Teams from Amirkabir University of Technology and Sharif University of Technology, both located in Tehran, tied for 17th place along with Penn State and the University of Illinois.[20] Some sixty teams from forty-one schools participated in the Tehran regionals leading up to the Asia-

Pacific regionals and then the world finals. Four of the top ten in the Tehran regionals were from Sharif University of Technology.[21] Sharif did even better in the 2006 contest, placing 13th, ahead of all US schools except the 8th-ranked Massachusetts Institute of Technology.[22] These results show that Iranian schools are producing the programming talent needed to conduct CNO, even if the skills are being employed for other purposes.

We identified several universities engaged in CNO education research. These include Sharif and Amirkabir, plus the University of Isfahan and Isfahan University of Technology.

At Sharif University of Technology, we found faculty and students with interests in computer security. One professor, Shahram Bakhtiari, has taught courses titled Cryptography and Network Security, Computers and Networks Security, and Systems and Networks Security. According to the course description for the third, "Students who take this course become familiar with methods of attack and the ways to protect systems and networks." His Website includes links to class presentations, including one on "Hacking Techniques" and one on "IP Security Flaws." Professor Bakhtiari has also published numerous papers on cryptography in journals and conference proceedings. He ran three workshops on information security in Iran: a 1999 workshop held in conjunction with the Computer Society of Iran's annual international conference, a second 1999 workshop held with the Iranian Conference on Electrical Engineering, and a 2001 workshop held with the International Internet and Electronic Cities Conference.[23]

Mohammad Abdollahi Azgomi, a PhD candidate at Sharif, wrote his master's thesis on network security and published papers on firewalls and other security topics.[24] Hashem Habibi, a master's student in software engineering working with "a huge number of other people" on network security, has links to security and hacking sites on his homepage at Sharif. His Website also has photos of himself and others associated with the Network Security Center and with "Seclab."[25] Sauleh S. Etemad, an alumnus of Sharif, taught courses and wrote technical reports on network and operating systems security at Iran's Advanced Information and Communication Technology Center before going on to earn a master's degree in electrical and computer engineering from Michigan State University. At Sharif, he completed his bachelor's thesis on operating systems security.[26]

In late 2005, I received an e-mail from a graduate student at Sharif who was completing a master's thesis on the topic of stream ciphers. His research interests included coding and cryptographic protocols, and he was interested in pursuing a PhD as a member of my group. He had already published two conference papers.

Sharif has hosted information security conferences, including the Second Iranian Society of Cryptology Conference and the Operating System and Security Conference 2003. In addition, it has hosted more general IT-related conferences and a conference on electronic warfare.[27]

Amirkabir University of Technology houses a Data Security Research Labo-

ratory within the department of Computer Engineering and Information Technology. The role of the laboratory is to help promote "research and innovations on
computer, information, and communications security" and to help train engineers and scientists in related areas.

Two students affiliated with the lab, Haamed Gheibi and Salman Niksefat,
taught a workshop on hacking operating systems at a conference held in Tehran
in 2004. They also posted information about a Microsoft Windows security flaw
on a computer security electronic mailing list, Bugtraq, in 2003, after unsuccess-
ful attempts to gain the attention of Microsoft. Gheibi represented Amirkabir in
the 2003 ACM programming contest.[28]

Several faculty members at Amirkabir listed computer security as an area of
interest. One professor, Mehran Soleiman Fallah, works extensively in the computer security field. His PhD thesis was on denial-of-service attacks, and he has
published several papers on this topic. He has also taught an undergraduate
course on network security and three graduate courses on information security
and network security.[29]

At the University of Isfahan, we found two professors who conduct computer
security research: Ahmad Baraani-Dastjerdi and Behrouz Tork Ladani.
Baraani's area of research includes cryptography, database security, and security
in computing.[30] Ladani's includes cryptographic protocols, information system
security, and network security. In 2005, Ladani also taught undergraduate
courses on cryptography and network security and on security in computer
systems. He received his PhD from the University of Tarbiat Modares, Iran,
where he wrote his thesis on cryptographic protocols.[31] This would suggest that
faculty at Tarbiat Modares are also conducting CNO-related research, which is
confirmed in the next paragraph.

Isfahan University of Technology hosted the Third Iranian Society of Cryptology Conference in September 2005. The conference covered a broad range of
topics in cryptography and computer and communications security. Several
faculty members at the university served on the conference committee, so we
can assume that there is some CNO-related research taking place at the school.
The committee also included representatives from Sharif University of Technology (nine people, including Bakhtiari), Amirkabir University of Technology
(Fallah), the University of Isfahan (two, including Baraani), Tarbiat Modares,
and several other schools and research institutions.[32] We did not attempt to track
down all thirty-four people on the committee, but the size of the committee
alone indicates a substantial community of security researchers in Iran, most
likely numbering at least a hundred or two. That Iran has a Society of Cryptology, which has sponsored at least three conferences, is further proof of an active
and established security research community.

We found several Websites in Farsi relating to network security. These
included sites for the IR Computer Emergency Response Team
(www.ircert.com), Iran Security (weblog.iransecurity.com), Iran Virus Database
(www.irvirus.com), and Hat-Squad Security Group (www.hatsquad.com). These
sites appear to discuss network vulnerabilities, with the objective of promoting

better security.[33] Hat-Squad offers security risk assessment, training, consultancy, incident response, penetration testing, and advisories that describe vulnerabilities and exploits.

We did not find any research or discussion on how Iran might employ CNA against its adversaries or the need to defend critical infrastructures in Iran from adversary CNA. The focus seems to be on security in general and on technology.

Government and foreign relations

The Iranian government promotes research and development in IT through several institutions, among them the Iran Telecommunications Research Center, the Technology Cooperation Office, Guilan Science and Technology Park, and Pardis Technology Park.

The Iran Telecommunications Research Center (ITRC) was formed in 1970 as the research arm of the Ministry of Information and Communications Technology. Research is organized into four departments: Information Technology, Strategic Management, Networking, and Transmission. Network security and security management are part of the center's research agenda, and one of the workshops on information security run by Shahram Bakhtiari of Sharif University of Technology was held at ITRC.[34]

ITRC is also involved in standards setting. It is a member of the European Technical Standards Institute and has created study groups aligned with the International Telecommunications Union study groups. Study group 17 is on security, languages, and telecommunications software.[35]

We found three researchers at the center who had presented papers on network security at international meetings. Mehdi Rasti, Davood Sarramy, and Mahmood Khaleghi gave a paper on network security assessment at a computer applications conference in Orlando, Florida, in 2004. In 2003, Rasti gave a paper on anomaly detection at the same conference in Las Vegas, Nevada.[36]

The Technology Cooperation Office (TCO) was founded in 1984 to serve the president of Iran. Its mission is to support development and cooperation in advanced technologies, including IT. Among the forms of support it offers Iranian institutions are coordinating joint research projects and establishing relations with foreign industrial and scientific research centers.[37]

Guilan Science and Technology Park was established in 1989 as the Iranian Research Organization of Science and Technology. The research center was reorganized as a technology park in 2002. One of the focus areas for the park is IT, and several IT companies have offices in the park. We did not identify any CNO-specific activity at the park.[38]

Pardis Technology Park (PTP), located 20 km from Tehran, was established in 2001 by TCO in order to create an environment for researchers, educators, and companies suitable for developing Iran's high-tech industry.[39] PTP's objectives are to intensify high-tech industry development; promote cooperation among industry, academia, and government research centers; create synergy between private and state sectors; commercialize know-how and innovations generated by research centers; and promote research and development in the private sector. PTP is run by a board of directors whose members are designated by TCO and Sharif University of Technology. The network security company Sharif Secure Ware is among the forty-five companies that have signed a contract to purchase land at the park.[40]

We found no evidence that the Iranian government was developing a CNA/E capability against its adversaries. However, given Iran's pursuit of asymmetric warfare capabilities, including nuclear weapons, ballistic missiles, and support for terrorism,[41] it is possible that it will pursue, if it is not already, a CNA/E capability as well. If so, it might collaborate with North Korea, which purportedly has been training cyber warriors for years (discussed below). According to reports, Iran has cooperated with North Korea on military technology training and transfer in the past, including development of missile systems. Iran has also sent military and intelligence officers to North Korea for training in psychological warfare and counterespionage.[42]

Hacking and cyber attacks

Iran has numerous hackers and hacking groups, some of which also sell network and security services. One such group is IHS Iran Hackers Sabotage. According to their Website, the group was formed in 2004 "with the aim of showing the world that Iranian hackers have something to say in the world wide security [sic]." After "rooting many important servers," they decided to participate in the "vulnerability assessment and exploitation process" and to offer a "highly secured hosting service." Their Website offers several original exploitation programs for download, each written for Visual C++ and based on vulnerabilities reported by others. The group consists of three active members, two of whom say they are university students.[43]

As of October 2005, IHS had defaced over 3,700 Websites.[44] All of the defacements we examined contained political messages. For example, a defacement on 25 July 2005 against the US Naval Station Guantánamo's public Website emphasized that Muslims were for peace, not terrorism, and that many had been harmed in Israel, Iraq, and Guantánamo.[45] On 2 October 2005, a defacement of a Novell site proclaimed that Iranians had a right to atomic energy and that "NO one can rule us not to use atomic power."[46]

Another group, the Ashiyane Digital Security Team, which sells Web hosting and network and security services, has defaced over 2,800 Websites. Their Website includes tools and tutorials on hacking and security, a discussion forum, a link to their Web defacements, and a list of over 3,500 registered users interested in security and hacking.[47]

Assuming most of the registered users are Iranian, which seems likely given that much of the Website is in Farsi, we can conclude that there are thousands of people in Iran interested in network security and hacking.

A defacement of the National Aeronautics and Space Administration's

Website on 11 August 2005 challenged US policy in the Middle East,[48] but most of the Ashiyane defacements we examined did not contain a political statement. In one case an attacker who goes by the name ActionSpider left his e-mail address and offered to help protect the site from other hackers; in another, the attacker offered free help patching the hacked server.

Ashiyane team members boast a wide range of experience in operating systems, programming languages, and hacking, including firewall penetration, database and operating system hacking, software cracking, and social engineering (conning a victim to perform some task, such as disclosing a password). Several members taught fee-based courses on hacking and other topics at a vocational school in Tehran.[49]

Among the other Iranian hacking groups we found are Iranian Boys Black Hat, Iran Hackers Association, Iran Babol-Hackers Security Team, Crouz Security Team, and Persian Crackers. Iranian Boys Black Hat has defaced as many sites as Ashiyane (over 2,800). As far as we could tell, none carried political messages. This was also true of defacements by Iran Babol-Hackers Security Team (over 400), which some members claim are "just for fun."

Iran formed a Defcon group in February 2004. Defcon groups are local groups associated with the annual Defcon meeting, which bills itself as the "largest underground hacking event in the world," drawing thousands of information security experts, hackers, and government officials to Las Vegas every summer for talks and hacking contests. The individual groups serve as local gathering places for discussions of technology and security. Iran's Defcon group was based in Tehran, but apparently it had ceased to exist by April 2006.[50]

Besides Web defacements, we found evidence of other political hacking within Iran. For example, the Weblog of former vice president Mohammad Ali Abtahi was hacked several times after he posted entries about government torture of other bloggers, and the Website of former presidential candidate Ali Larijani was subjected to a distributed denial-of-service attack. Larijani's campaign committee claimed that his site was hacked by the opposition. Bloggers theorized that the government was responsible for the attacks against Abtahi and Larijani, but no supporting evidence was provided.[51]

Iranians have also acquired and used software that bypasses the government's Internet filters. In an interview with Shift.com, Oxblood Ruffin, founder of Hacktivismo, reported that their software was being used in Iran. We did not find evidence of Iranians developing their own anticensorship software.[52]

Summary

Although less than 11 percent of Iranians were online at the end of 2005, Iran has a sizable community of interest and expertise in computer network attack and defense. We estimate that there are 100 or more academics working in information security, publishing research papers in journals and conference proceedings, hosting and attending conferences, and teaching courses on network security topics. Although we could not determine sponsorship for this work, it is probably fair to assume that it is at least approved, if not also funded, by the government.

We also estimate that there are thousands of additional hackers and network security specialists. Many of them have experience in breaking into Websites and conducting other types of attacks. Some offer network security products, services, and training.

The Iranian government is actively promoting many areas of IT, including networks, with the goal of stimulating economic growth. Although we found government-sponsored research in network security taking place within government labs, we did not identify any government involvement in cyber attacks or any government effort to develop a CNA/E capability against adversary countries. However, should government officials decide to develop such a capability, they could draw on the Iranian IT community to put together an attack team.

All of these findings indicate that Iran is concerned about network security and taking steps to defend its networks, advance the common body of knowledge in security, and exploit the commercial market for network security products and services. It also has its share of hackers, including people who deface Websites. This is all to be expected in today's interconnected world, which has been attracting an ever increasing body of cyber vandals, crooks, and spies, as well as people devoted to improving computer defense. Any country that would ignore network security would do so at its peril. From open sources, we did not find indications that Iran's efforts in network security are motivated by a desire to conduct crippling attacks against the infrastructures of other countries.

North Korea

Our study of the Democratic People's Republic of Korea (DPRK) was conducted by Navy Lt. Christopher Brown and completed in September 2004.[53] We found very little information coming directly from inside North Korea, and most of that was posted on Websites belonging to the government. Hence, we relied more on second-hand information provided by governments, news agencies, and scholars residing in other countries. The following subsections summarize some of the key findings.

IT industry and infrastructure

North Korea is one of the most disconnected countries in the world. In 2001, it had 1.1 million telephone lines,[54] which represents less than five lines per 100 population, compared with twenty-seven for Iran and 117 for South Korea and the United States. North Korea began to develop a cellular infrastructure, but in May 2004 the government banned mobile phones in order to limit foreign influences. The country owns two satellites, an International Telecommunications Satellite (Intelsat) and a Russian satellite, both operating in the region of the Indian Ocean. The French provide technical support.[55]

The situation with the Internet appears to be even worse. Although North

Korea has a top-level domain name (.kp) and two assigned Class C Internet protocol (IP) address blocks with 131,072 addresses, we found no evidence of any activity originating from these assigned IP addresses or the .kp domain.[56] A Google search of the .kp domain returned 147 hits on 24 October 2005, but none of the Websites were accessible, and no content was displayed with the search results, unlike most searches, which return two lines of content for each matching Website. It is possible that the sites are registered but not yet used. Alternatively, the sites may be up but inaccessible from the United States or outside North Korea.

We did find North Korean Websites hosted in other countries, including China, Japan, and Australia. The small handful of official state-sponsored sites we found were located on servers in China and Japan.[57] The Website for the Korean Central News Agency of DPRK, for example, is in Japan (at www.kcna.co.jp/).

Internet access in North Korea is extremely limited. An Internet café was opened in Pyongyang in May 2002, but the rates were reported to be about $10 per hour, more than one-fifth of the average North Korean's monthly earnings. Thus, the café is believed to serve mainly visiting businessmen, tourists, and diplomats. Some hotels in Pyongyang also provide Internet access, but again for visitors.[58] We did not find any information regarding Internet access for the general population. Considering the ban on cell phones, it seems likely that Internet access is highly restricted, if even available.

North Korea has a national intranet. The Kwang Myong (Bright Star) Network runs through fiber-optic cable with a backbone capacity of 2.5 gigabytes per second.[59] Developed in 1996 with the goal of linking various research and academic institutions, the Bright Star Network now also includes government and military agencies, as well as public access. By November 2004, several PC cafés were open in Pyongyang, providing access to e-mail, internal Websites, chat, online games, and streaming movies over a 100 megabit-per-second fiber-optic link to the national intranet. The largest café, located by a subway station, has around 100 computers.[60] A 2001 report indicated that North Korea had begun testing a firewall between the Bright Star Network and the Internet in order to screen and restrict information flows in both directions.[61]

Telecommunications and networking depend on power, and North Korea's electrical infrastructure is both antiquated and unreliable, with frequent power outages and poor frequency control. Since reliable and stable power is needed for sustained computer network operations, North Korea's ability to conduct CNA/E against its adversaries is probably limited.[62]

North Korea has developed a personal data assistant (PDA), the Hana-21, based on an original Korean operating system. However, much of its IT hardware sector is technologically dated, and computers and communications equipment are imported from China and Southeast Asia.[63] Technology exports to North Korea are severely restricted under the Wassenaar Arrangement on Export Controls for Conventional Arms and Dual-Use Goods and Technologies, limiting North Korea's ability to acquire advanced information technologies from signatories of the treaty (which includes the United States and South Korea but not China).

North Korea's software industry is closely tied into its research institutions, including the Korean Computer Center, the Pyongyang Programming Center, and Kim Il Sung University. Areas of focus include voice recognition, language translation, gaming, animation, multimedia, and biometrics.[64] Except for biometrics, which can be used for network security, these technologies are not germane to CNO.

We did not find any laws specifically addressing the Internet, including computer crime laws. However, telecommunications are heavily censored, and all international telephone calls are facilitated through a state-run exchange operator, which is closely monitored. Until computers, telephones, and the Internet become more prevalent, North Korea may not see much need for computer crime laws.

Academic and research community

North Korean leader Kim Jong Il has said that there are three basic types of fools in the twenty-first century: people who smoke, people who do not appreciate music, and people who cannot use the computer. An avid Internet user, he has stated that IT is the future of North Korea and that those who do not educate themselves in it will be left behind. Hence, it is not surprising that computer education is mandatory and emphasized, starting in grade school. Computer science has topped the list of curriculum choices among young military officers and college students, and possessing a computer-related job is considered a sign of privilege. North Korea does not participate in the ACM programming contest, but students can submit software they have developed to a government-sponsored national programming contest.[65]

We found three major academic institutions in North Korea actively involved in IT: Pyongyang University of Computer Technology, Kim Chaek University of Technology (KUT), and Kim Il Sung University. Faculty at Kim Il Sung University have developed security-related software products, including Worluf Anti-Virus and Intelligent Locker.[66]

In 2001, KUT and Syracuse University began discussions on the possibility of research collaboration in integrated information technology. By June 2004, KUT representatives had made three visits to Syracuse, and the Syracuse team had made one trip to North Korea. The general area of collaboration is systems assurance, in particular technology to foster trusted communications. Although "trusted communications" is often linked with cryptography and network security, the group seems to be concerned more with integrity, safety, and reliability than network defense. The current focus has been on using open-source software to produce a back-end library management system for the KUT digital library. The group has produced designs for twin research labs, software specifications, joint work on proving software correctness, research presentations, and an academic paper.[67]

Government and foreign relations

North Korea has seven research institutions focused on IT. The most prominent are the Pyongyang Informatics Center, the Korea Computer Center, the DPRK Academy of Sciences, and Silver Star Laboratories.

The Pyongyang Informatics Center (PIC) was established in 1986 to develop computer-based management techniques and to help promote the use of computers in government and industry. Its primary focus is software development, and PIC has produced a variety of products, including the software filters used between the Bright Star Network and the Internet, which serve a role in computer network defense as well as censorship. However, most of PIC's development work seems to be in areas unrelated to CNO, including electronic publication, computer-aided design, embedded Linux, Web applications, interactive programs, accounting, and virtual reality. This assessment is supported by a report that in 2001, researchers at PIC requested 250 IT books from South Korea; they were especially interested in books on graphics and virtual animation but also on common operating systems and communication methods. The list did not include any books relating to cyber security.[68]

The Korea Computer Center (KCC) was established in 1990 to promote computerization. With 800 employees at its inception, it has produced some of North Korea's cutting-edge software, including systems for voice recognition, fingerprint identification, and artificial intelligence. It has produced a Korean version of the Linux operating system, and its chess playing software has dominated Japan's annual Chinese chess competition.[69]

The KCC is directed by Kim Jong Nam, the son of Kim Jong Il. Nam, who also heads the State Security Agency (SSA), which is North Korea's intelligence service, moved SSA's overseas intelligence unit into the KCC, according to a South Korean newspaper. South Korean media have also claimed that the KCC is "nothing less than the command center for Pyongyang's cyber warfare industry, masquerading as an innocuous, computer geek-filled software-research facility."[70]

The DPRK Academy of Sciences and Silver Star Laboratories are also involved in software development. Between them, they have produced software for language translation, optical character recognition, artificial intelligence, multimedia, remote control, and communications. We did not find any indications that either institute had developed CNO-related software.[71]

In 1984, North Korea established the Mirim Academy, which offered a two-year program in IT and electronic warfare for top military students. Two years later, the school became a five-year college, Mirim College, and opened admissions to high school students from the top percentile. The school, also known as the Automated Warfare Institute, purportedly offers curricula in command automation, computers, programming, automated reconnaissance, and electronic warfare.[72]

According to a June 2003 news report, Maj. Gen. Song Young-keun, commanding general of South Korea's Defense Security Command, said that North

Korea has been producing 100 cyber soldiers annually.[73] In May 2004, at a conference in Seoul organized by the Korea Information Security Agency (KISA), Song said that "Following orders from Chairman Kim Jong Il, North Korea has been operating a crack unit specializing in computer hacking and strengthening its cyber-terror ability." He said that the hackers were handpicked from among the top graduates of Kim Il Sung Military Academy and given intensive training in computer-related skills before being assigned to the hacker's unit.[74] According to East-Asia-Intel.com, which provides news on the Far East, Mirim College was renamed Kim Il Military Academy and later Pyongyang College. The news site also reported that Byun Jae-Jeong, a research fellow at the South Korea Agency for Defense Development, claimed that the cyber agents had a technical ability on a par with that of CIA hackers and that they were able to "infiltrate and gather information from Web servers from various countries."[75]

Developing a CNA/E capability is certainly consistent with Kim Jong Il's interest in IT and his military objectives – to "disturb the coherence of South Korean defenses in depth including its key command, control, and communications, and intelligence infrastructure."[76] Moreover, Richard Clarke, former special adviser to the president for cyberspace security, reported that North Korea was "developing information warfare units, either in their military, or in their intelligence services, or both."[77]

Hacking and cyber attacks

At the 2004 conference in Seoul, Maj. Gen. Song Young-keun claimed that North Korea's military hackers had been breaking into the computer networks of South Korean government agencies and research institutes to steal classified information.[78] We also found reports of other cyber attacks being attributed to North Korean hackers. However, Director Baek, of South Korea's National Intelligence Service (NIS), told us in a telephone interview in April 2004 that NIS had no knowledge of confirmed CNA/E activities originating from within North Korea, or of North Korea sponsoring CNA/E against any country. This view was echoed by officials at KISA.[79]

We found no evidence that North Korea has hackers operating outside the government. Given the severe restrictions on Internet use within the country, any hacking being conducted from North Korea would most likely be government sponsored. Within the government, hacking seems to be confined to the KCC and Mirim College (Kim Il Military Academy/Pyongyang College).

Summary

North Korea most likely has a CNO capability within its military and intelligence services. It appears to recognize the value of IT and CNO to its future and to have devoted resources to training and supporting cyber warfare units.

Whether North Korea's CNO capability has been used to attack targets in South Korea, the United States, or elsewhere is less certain. The capability may be used primarily for defensive purposes or for intelligence collection against foreign governments and businesses. However, if North Korean hackers are able to stealthily penetrate or exploit computer networks in order to acquire secrets, they could as well use their skills to damage or disrupt these networks.

North Korea faces several obstacles to developing and deploying an advanced CNO capability. Its highly restricted Internet connectivity and unreliable and antiquated electrical infrastructure could interfere with the conduct of attacks, especially sustained attacks. Trade restrictions make it difficult for the country to acquire the latest hardware and software platforms, which in turn hampers its ability to develop and test attacks against these systems. Restrictions on Internet access would make it hard for North Korea to acquire hacking tools and information from the Internet, and to use or build on the work of tens of thousands of others in the world. Because much of the North Korea's IT research and development effort is in areas unrelated to CNO, the country's own academic and public communities would have little to offer in the way of CNO expertise. Internet restrictions would also preclude North Korean youth from getting involved in the Internet hacking scene and building up knowledge and skills that could later be channeled into government-sponsored activity. CNO agents would have to be trained from scratch.

While these hurdles do not imply that North Korea could not develop a powerful CNO capability, they suggest that a certain amount of skepticism may be appropriate when assessing claims about the effectiveness of that capability.

Conclusions

Our study concluded that both Iran and North Korea have a CNO capability. However, whereas the capability we identified for Iran lies within its academic and research communities and the general population, North Korea's lies mainly within its military.

We did not find evidence that either country had a highly sophisticated capability that would even come close to matching that in many other countries, including Australia, China, Russia, the United Kingdom, and the United States. Both countries, but especially North Korea, operate at a disadvantage because of trade restrictions prohibiting exports of advanced Western technologies to them. North Korea's disadvantage is compounded by its extreme isolation, not just from the Internet but from most of the world. Iran is plugged into the Internet and the international business, security, and hacking communities and thus can better leverage technologies and knowledge developed outside its borders. Moreover, the Iranian government can build on the knowledge and skills of its own population as participants in these international communities. North Korea is confined to whatever CNO capability it can develop in-house, behind government doors.

There are several limitations to our study. First, and perhaps most important, it is difficult and risky to draw conclusions based on a lack of evidence. It could be that both countries have highly advanced CNO capabilities, and that we just did not look hard enough or in the right places. As noted earlier, we did not have access to government officials in either country, and we did not use classified information from our own intelligence services, which no doubt limited what we could learn, especially about military capabilities. Our limited resources – we could not conduct every possible Internet and library search, follow every link, and translate every foreign Website and document – also limited our data collection.

Another limitation is that our assessment is mainly qualitative. We attempted to measure a few factors, including the number of security researchers and hackers in a country, the percentage of the population with Internet access, and the size of the security industry, but we did not formulate specific metrics that would allow one to rate a country's CNO capability on, say, a scale from 0 to 10. That said, I might rate Iran at 2 or 3 and North Korea at 1. I would rate Iran higher if we had evidence of a strong CNO capability within its military.

A third limitation is that our research and assessments were inherently subjective, biased by our own preferences and beliefs. These included beliefs that it would be difficult to develop a strong CNO capability in isolation and that a CNO capability within a country's population could be leveraged by a government to develop or strengthen its own.

These limitations present an opportunity for future research. Currently, however, we have shifted our focus to the terrorist threat. We have developed a methodology to assess the CNO threat of terrorists, in the process applying it to al-Qaeda and the global jihadists.

Notes

1 D. E. Denning, "Activism, Hacktivism, and Cyberterrorism: The Internet as a Tool for Influencing Foreign Policy," chap. 8 in J. Arquilla and D. Ronfeldt (eds), *Networks and Netwars*, Santa Monica: RAND, 2001, pp. 276–7.
2 D. E. Denning, "Information Technology and Security," chap. 4 in M. E. Brown (ed.), *Grave New World*, Washington, DC: Georgetown University Press, 2003, p. 98.
3 J. P. Patterson and M. N. Smith, "Developing a Reliable Methodology for Assessing the Computer Network Operations Threat of Iran," master's thesis, Naval Postgraduate School, September 2005.
4 C. Brown, "Developing a Reliable Methodology for Assessing the Computer Network Operations Threat of North Korea," master's thesis, Naval Postgraduate School, September 2004.
5 The definitions here are based on *Field Manual (FM) 3–13, Information Operations: Doctrine, Tactics, Techniques, and Procedures*, US Army, November 2003. They are consistent with revisions to the Joint Doctrine for Information Operations, Joint Pub (JP) 3–13, Joint Chiefs of Staff.
6 C. Billo and W. Chang, *Cyber Warfare: An Analysis of the Means and Motivations of Selected Nation States*, Institute for Security Technology Studies, November 2004, www.ists.dartmouth.edu/directors-office/cyberwarfare.pdf (accessed 26 April 2006).
7 The Website for the contest is at icpc.baylor.edu/icpc/ (accessed 26 April 2006).
8 Patterson and Smith, "Developing a Reliable Methodology for Assessing the Computer Network Operations Threat of Iran."
9 Ibid.
10 Ibid.
11 Ibid.

12 Ibid.
13 Information technology statistics for 2004, International Telecommunications Union, www.itu.int/ITU-D/ict/statistics/at_glance/Internet04.pdf (accessed 13 October 2005).
14 Internet World Stats, www.internetworldstats.com/stats5.htm (accessed 13 April 2006).
15 Internet Filtering in Iran in 2004–2005, OpenNet Initiative, www.opennetinitiative. net/studies/iran/ONI_Country_Study_Iran.pdf (accessed 13 October 2005).
16 Patterson and Smith, "Developing a Reliable Methodology for Assessing the Computer Network Operations Threat of Iran."
17 Website of Sharif Secure Ware, www.amnafzar.com/ (accessed 17 October 2005).
18 "Leading Iranian Software Company System Group Made Exclusive Contract with Securepoint Security Solutions," 19 July 2005, www.securepoint.cc/press_ partnership_iran.html (accessed 17 October 2005).
19 Patterson and Smith, "Developing a Reliable Methodology for Assessing the Computer Network Operations Threat of Iran."
20 ACM programming contest rankings for 2005, icpc.baylor.edu/past/icpc2005/ finals/Standings.html (accessed 14 October 2005).
21 Twenty-ninth ACM International Collegiate Programming Contest, 6th Asian Regional Contest in Iran, sharif.ir/~acmicpc/acmicpc04/index.html (accessed 14 October 2005).
22 ACM programming contest rankings for 2006, icpc.baylor.edu/icpc/Finals/ default.htm (accessed 26 April 2006).
23 Website of S. Bakhtiari, sharif.ir/~shahram/ (accessed 14 October 2005).
24 Website of M. Azgomi, mehr.sharif.edu/~azgomi/ (accessed 14 October 2005).
25 Website of H. Habibi, ce.sharif.edu/~hhabibi/ (accessed 14 October 2005).
26 Résumé of S. Etemad, www.egr.msu.edu/~etemadys/Resume.pdf (accessed 14 October 2005).
27 Research Website of Sharif University of Technology, www.sharif.ir/en/ research/ (accessed 2 November 2005).
28 Patterson and Smith, "Developing a Reliable Methodology for Assessing the Computer Network Operations Threat of Iran."
29 Website of M. Fallah, ce.aut.ac.ir/~fallah/ (accessed 14 October 2005).
30 Website of A. Baraani-Dastjerdi, eng.ui.ac.ir/ahmadb/ (accessed 14 October 2005).
31 Website of B. Ladani, eng.ui.ac.ir/ladani/ (accessed 14 October 2005).
32 Website of Third Iranian Society of Cryptology Conference, iscc2005.org/ iscc/index.php?sel_lang=english (accessed 14 October 2005).
33 Patterson and Smith, "Developing a Reliable Methodology for Assessing the Computer Network Operations Threat of Iran."
34 Ibid.
35 Ibid.
36 Ibid.
37 Ibid.
38 Ibid.
39 Website of Pardis Technology Park, www.hitechpark.com/ (accessed 17 October 2005).
40 Ibid.
41 Bill Gertz, "Iran Militants in Power Stir Fears," *Washington Times*, 14 October 2005, www.washingtontimes.com/national/20051013–114716–3258r.htm   (accessed   17 October 2005).
42 Patterson and Smith, "Developing a Reliable Methodology for Assessing the Computer Network Operations Threat of Iran."
43 Website of Iran Hackers Sabotage, www.ihsteam.com (accessed 18 October 2005).
44 The statistics on Web defacements were obtained 18 October 2005 from the Zone-H Website at www.zone-h.org. The site also hosts mirrors of defacements.
45 Zone-H mirror of Web defacement, www.zone-h.org/en/defacements/mirror/ id=2645159/ (accessed 20 October 2005).

46 Zone-H mirror of Web defacement, www.zone-h.org/en/defacements/mirror/id=2917409 (accessed 18 October 2005).
47 Website of Ashiyane Digital Security Team, www.ashiyane.com/ (accessed 18 October 2005).
48 Zone-H mirror of Web defacement, www.zone-h.org/en/defacements/mirror/id=2757538/ (accessed 20 October 2005).
49 Patterson and Smith, "Developing a Reliable Methodology for Assessing the Computer Network Operations Threat of Iran."
50 Website of Defcon Groups, www.defcon.org/html/defcon-groups/dc-groups-index.html (accessed 15 October 2005 and 26 April 2006).
51 Patterson and Smith, "Developing a Reliable Methodology for Assessing the Computer Network Operations Threat of Iran."
52 Ibid.
53 Brown, "Developing a Reliable Methodology for Assessing the Computer Network Operations Threat of North Korea."
54 The Central Intelligence Agency, The World Factbook, North Korea, www.odci.gov/cia/publications/factbook/geos/kn.html (accessed 20 October 2005).
55 Brown, "Developing a Reliable Methodology for Assessing the Computer Network Operations Threat of North Korea."
56 Ibid.
57 Ibid.
58 Ibid.
59 Ibid.
60 "'PC Café' attracts Youth in Pyongyang," December 2004, www.vuw.ac.nz/~caplabtb/dprk/NK_S&T.htm (accessed 21 October 2005).
61 Brown, "Developing a Reliable Methodology for Assessing the Computer Network Operations Threat of North Korea."
62 Ibid.
63 Ibid.
64 Ibid.
65 Ibid.
66 Ibid.
67 S. T. Song *et al.*, "Bilateral Research Collaboration Between Kim Chaek University of Technology (DPRK) and Syracuse University (US) in the Area of Integrated Information Technology," Prepared for the Asian Studies on the Pacific Coast (ASPAC) 2003 Annual Meeting, Honolulu, Hawaii, www.koreasociety.org/FYI/20030630-ASPAC-Kim-Chaek-Syracuse-rv.pdf (accessed 21 October 2005); "Project Status Report: July 2004 on The KUT/SU Research Collaboration," Nautilus Institute, DPRK Briefing Book, www.nautilus.org/DPRKBriefingBook/economy/30-KoreaSociety.html (accessed 21 October 2005).
68 Brown, "Developing a Reliable Methodology for Assessing the Computer Network Operations Threat of North Korea."
69 Ibid.
70 J. Larkin, "North Korea Preparing for Cyberwar," *Far Eastern Economic Review*, 25 October 2001, archive.infopeace.de/msg00464.html (accessed 24 October 2005).
71 Brown, "Developing a Reliable Methodology for Assessing the Computer Network Operations Threat of North Korea."
72 Ibid.
73 "North Korea Suspected of Training Hackers," Associated Press, June 2003, smh.com.au/articles/2003/06/10/1055010959349.html (accessed 24 October 2005).
74 AFP, "North Korean Military Hackers Unleash 'Cyber-Terror' on South Korean Computers," 27 May 2004, www.freerepublic.com/focus/f-news/1143440/posts (accessed 24 October 2005).

75 "North Korea's Cyber Guerrillas Called CIA-Class Threat to US Pacific Command," East-Asia-Intel.com, 7 June 2005.
76 Brown, "Developing a Reliable Methodology for Assessing the Computer Network Operations Threat of North Korea."
77 Ibid.
78 AFP, "North Korean Military Hackers Unleash 'Cyber-Terror' on South Korean Computers."
79 Brown, "Developing a Reliable Methodology for Assessing the Computer Network Operations Threat of North Korea."

9　Blogs and military information strategy

James Kinniburgh and Dorothy Denning

In September 2004, with the presidential campaign in full swing, the producers of the CBS television news show *60 Minutes Wednesday* received a memo purporting to show that the sitting president, George W. Bush, had used his family connections to avoid his military service obligations as a National Guardsman in the early 1970s. Given the controversy and ratings the story would generate, it was just too good not to run. On cursory inspection, the documents and their source appeared legitimate. *60 Minutes* producer Mary Mapes and anchorman Dan Rather decided to air it.

Within minutes after air time, discussion participants at the conservative Website FreeRepublic.com asserted that the documents were faked. Bloggers at Power Line[1] and Little Green Footballs (littlegreenfootballs.com) soon picked up these comments and posted them and their associated hyperlinks on their own blogs. The main clues that the now infamous Killian memos were forgeries, the bloggers pointed out, were the superscripted "th" and the Times New Roman font; both indicated the use of modern word-processing programs rather than a 1972-era typewriter. The signatures on at least two of the documents appeared to have been forged, and some commentators with military experience called into question the very format of the memos. The story was given even greater attention after Internet journalist Matt Drudge posted a link to FreeRepublic thread on his own Website, *The Drudge Report* (www.drudgereport.com).

What followed initially was what is known as a "blogswarm," where the story was carried on multiple blogs, and then later a "mediaswarm." With the attention of bloggers, the media, and the public fixed on "Memogate" (one of the incident's popular monikers) and CBS unable to authenticate the documents, several CBS employees, including producer Mary Mapes, were asked to resign. Within a month, Dan Rather announced his own retirement.

What garnered considerable interest afterward was how a group of nonprofessional journalists was able to outperform and bring down two icons of the traditional media, CBS and Dan Rather. Former CBS News executive Jonathan Klein said of the bloggers, "You couldn't have a starker contrast between the multiple layers of checks and balances [at *60 Minutes*] and a guy sitting in his living room in his pajamas writing."[2]

Some columnists, such as Corey Pein at the *Colombia Journalism Review*, explained the spread of the story as the result of journalistic haste and the rapid coalescence of popular opinion, supported and enhanced by a blogging network of Republican story spinners.[3]

CBS, in its final report on the matter, offered its own explanations for the problems surrounding the story. The CBS reviewers identified four major factors that contributed to the incident: weak or cursory efforts to establish the documents' source and credibility, failed efforts to determine the document's authenticity, nominal efforts at provenance, and excessive competitive zeal (the rush to air).[4]

Although the initial questions about the CBS story were posted on a discussion forum instead of a blog, the partially erroneous attribution of the "Memogate" story – and other stories that followed – to bloggers no doubt increased public awareness of blogs and blogging as well as the potential they have to influence opinion and events. Governments have noticed this potential, and many authoritarian governments censor blogs they perceive as a threat to their regimes. Iran has imprisoned bloggers who offended the ruling mullahs. At the same time, however, Iranian officials have recognized the value of blogs to information strategy, and in February 2006 the regime held the Revolutionary Bloggers Conference to promote pro-regime blogs.[5]

The rise of military bloggers from deployed areas such as Iraq has raised concerns among US Department of Defense officials that information posted on a blog could compromise operations security. *Stars and Stripes*, a newspaper that caters to overseas military personnel, quoted a recent memo from the Army Chief of Staff, General Peter Schoomaker:

> The enemy aggressively "reads" our open source and continues to exploit such information for use against our forces.... Some soldiers continue to post sensitive information to Internet Web sites and blogs.... Such OPSEC [operations security] violations needlessly place lives at risk and degrade the effectiveness of our operations.[6]

In this chapter we explore the possibility of incorporating blogs and blogging into military information strategy, primarily as a tool for influence but also for gathering intelligence. To that end, we examine the value of blogs as targets of, or platforms for, military influence operations and supporting intelligence operations. Influence operations are a subset of information operations (IO) that includes the core capabilities of psychological operations (PSYOP) and military deception, along with the related capabilities of public affairs, military support to public diplomacy and civil affairs/civil–military operations.

To evaluate the IO potential for blogs, we seek to answer three questions:

1   Are blogs truly influential, and if so, in what manner?
2   Does the information environment support blogging as part of an information campaign?
3   Can blogs provide a significant source of intelligence or measures of effectiveness for influence operations?

Before addressing these questions, however, we review the nature and structure of the blogosphere.

Blogs and the blogosphere

The blogosphere is the world of blogs (short for Weblogs), bloggers, and their interconnections. It lies mostly within the Web, but it intersects traditional media and social networks as well. For example, the *Washington Post* features several blogs on its Website alongside the newspaper's traditional editorials and op–ed columns, and blog entries often make their way into social networks through e-mail.

Entrenched inequality and hierarchy

The Web is generally considered egalitarian; anyone can set up a Website, and anyone can access the Web. In practice, however, not all nodes of the Web are equal. Indeed, the network formed by Web pages and their links to each other is often said to follow a power law, meaning that the probability that any given node (a Web page) is linked to k other nodes is proportional to $1/k^n$ for some constant n (about 2 for the Web). The power law implies that for any given node in the network, the odds are that it will have relatively few connections. A few favored nodes, however, will have a disproportionately large number of connections. A network that follows a power law distribution is said to be scale free.[7]

Research has shown that the blogosphere itself approximates a scale-free net. This is illustrated in Figure 9.1, which shows the middle part of the curve as

Technorati™

Blogs by Authority (as measured by links from unique sources)

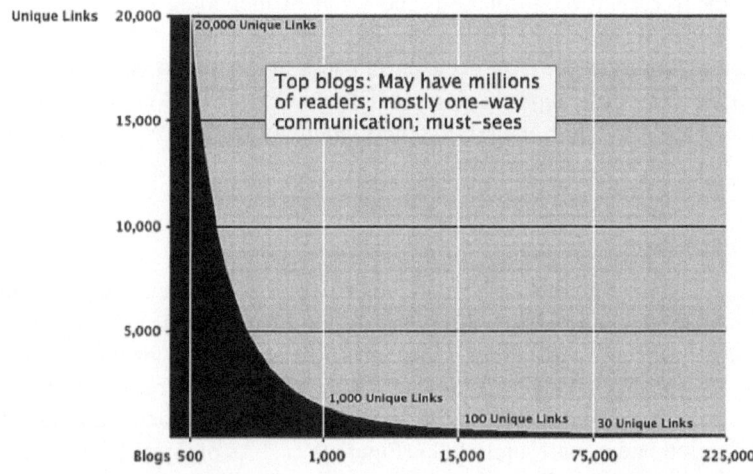

Figure 9.1 Number of blogs by number of unique links (source: Technorati[10]).

computed by Technorati in 2006.[8] However, the power law may not be an exact fit to the blogosphere. Daniel Drezner and Henry Farrell found that the distribution of links among political blogs was closer to a lognormal distribution rather than a straight power law.[9]

Regardless of whether the power law or lognormal distribution better models the blogosphere, both distributions are highly skewed. This inequality reflects a hierarchy among blogs. Simply put, a few blogs are extremely popular – they have large numbers of inbound connections – while the majority are hardly noticed or lie somewhere in between. This hierarchy is generally reinforced through a process called "preferential attachment." As a scale-free network grows, the established nodes that are already highly connected tend to pick up the new links. In short, "the rich get richer." The nodes with few connections also gain links, but the overall inequality is reinforced.[11] This does not imply that new nodes cannot rise to prominence or that established ones will never lose their position, only that change is likely to be slow and that the overall shape of the distribution will persist. Drezner and Farrell suggest that the lognormal distribution is more suited to change than the power law, as it allows poorly linked nodes to increase their ratings.[12]

The implications of the entrenched inequality of the blogosphere for influence operations are threefold. First, other things being equal, the blogs to which other Web pages link the most often are likely to be among the most influential. Second, the vast majority of blogs can be ignored, which allows certain IO efforts to be concentrated on the most popular blogs. Third, bringing a new blog into prominence is likely to be a slow and difficult process. It is not impossible, however. Technorati, which ranks blogs by number of links from unique sources, found that the third and fifth ranked blogs in February 2006 were not even among the top 100 in April 2005.[13]

It is also useful to consider that the preferential attachment reflected in the growth of the Web and blogosphere is the result of individual users' choices. Links are added by people, not some abstract network entity implementing a formula for network growth. The decision to add a link is based on social considerations, such as trust, reputation, or social pressure, as well as on personal preferences. Thus, the forces underlying the structure of the Internet and the blogosphere are ecological and selective.

Who blogs and why

According to the Pew Internet and American Life Project, between February and November 2004, the numbers of blog creators and blog readers both increased by at least 50 percent, with about 7 percent of Internet users having created blogs and 27 percent having read blogs.[14] More recent data from Technorati indicate that the blogosphere doubled in size about every six months during the three-year period ending in March 2006 (Figure 9.2).[15] As of April 2006, Technorati was tracking 35.3 million blogs and recording 75,000 new ones each day. Some of these sites are not actual blogs but "splogs" (spam blogs), which mimic

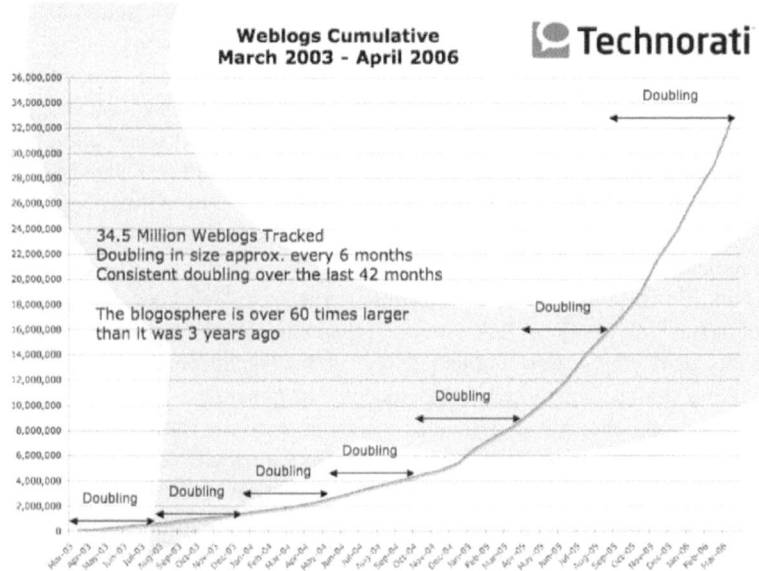

Figure 9.2 Growth of blogosphere (source: Technorati[17]).

blogs but contain entries full of hyperlinked text that directs readers to other sites. From 2 percent to 8 percent of the blogs created daily in late 2005 were said to fall into this category.[16]

The vast numbers and rapid growth of blogs should be interpreted as indicative of blogging's popularity as a format for information dissemination, not as a measure of blog influence – just as reporting the number of PSYOP leaflets dropped says nothing about their effectiveness in influencing a target audience.

The bloggers themselves generally match the demographic profile of American Internet users. Although the first bloggers tended to be young, male, college educated, and generally in or from the middle to upper middle income brackets,[18] since 2000, bloggers are just as often female as male and represent most minority groups and nearly every income bracket and educational level. Despite this demographic explosion, the majority of serious bloggers still tend to be less than fifty years of age, have a broadband connection to the internet and have at least some years' experience online.[19]

A study by Nardi *et al.* identifies five major motivations for blogging:[20]

- Documenting the author's life and experiences.
- Expressing opinions and commentary.
- Venting strong emotions.
- Working out ideas through writing (as MIT social scientist Sherry Turkle put it, using the blog and computer as "objects to think with"[21]).
- Forming and/or maintaining virtual communities.

Another motivation may be the potential for financial gain, as Internet advertisers seek to capitalize on bloggers' status and reputation by placing static, banner, and pop-up ads on their sites. Moreover, the number of professional bloggers is likely to increase as the media come to accept them as legitimate "citizen journalists." It is in this role of citizen journalist and media watchdog that bloggers such as Glen Reynolds, creator of the highly hyperlinked blog InstaPundit.com, revel. Precisely because of a perception that the big media are motivated less by the pursuit of truth and more by profit, corporate interests, and political agendas, many bloggers have cultivated reputations as vigorous challengers of the mainstream media.

InstaPundit's Reynolds is more than just a successful blogger. He is a law professor at the University of Tennessee and the author of several books and articles on political ethics, environmental law, and advanced technology. These credentials no doubt contributed to his success. In general, a blogger's objectives; qualifications and life experiences; skills in writing, framing arguments, and making use of the Web-page medium; personal attributes such as integrity; networks of personal contacts; and levels of interaction with the audience all contribute to the audience's assessment of the merit and credibility of his or her blog. These qualities are communicated to the audience through the blog, establishing the writer's online persona and reputation. Influence, therefore, starts with the characteristics of the blogger. According to researcher Kathy Gill, of the University of Washington, the most influential blogs were generally written by professionals with excellent writing skills.[22] Just as during World War II the military recruited the top Hollywood directors and studios to produce films about the war – in effect conducting domestic influence campaigns in the name of maintaining national morale and support for the war effort – waging the war against terrorism and its underlying causes, as spelled out in the National Security Strategy, may require recruiting the prominent among the digirati (probably those native to the target region) to help in any Web-based campaign.

The importance of credibility and reputation to a blog's influence must be taken into account when considering using a blog as a vehicle for information operations. This is especially critical given the poor image and reputation of the US government in many of the countries we want to influence.

Blog influence

Examples of the increased influence of blogs over the past two years include not only "Memogate" but also "Gannongate" and "Jordangate."

When the White House invited Jeff Gannon (pseudonym of James Gale Guckert), a blogger known to be sympathetic to the Bush administration, to join the White House press corps in 2004, the major media vilified him as an administration stooge and portrayed his inclusion in the corps as an attempt by the president to subvert the free press. Subsequent media scrutiny of Guckert revealed his lack of any "real" journalistic background and various sordid personal activities. The attention and negative publicity forced Guckert to retire from running his blog "Talon News" in February 2005.[23] Nevertheless, the White House's reaching out to bloggers brought credibility to blogging and helped thrust the phenomenon into mainstream awareness.

Military bloggers are credited with the fall of Eason Jordan, a senior executive at CNN who resigned in February 2005 over his remarks that US troops were deliberately targeting reporters in Iraq.

Nardi *et al.* cite two other examples of the political influence of blogs: Senator Trent Lott's loss of office after he gave tribute to Senator Strom Thurmond's mid-century segregationist run for the presidency and Howard Dean's success at fund-raising in the 2004 presidential campaign.[24] The major question that arises from these cases is how could (to paraphrase CBS's Klein) "some guys in pajamas writing at home" succeed in influencing not just the careers of prominent journalists, producers, and media executives but also potentially the course of a national election or public opinion about a war?

As noted earlier, blog influence can be affected by the structure of the blogosphere, and particularly by the network of hyperlinks connecting one blog to another. To illustrate, imagine starting at any random blog. By following a series of links from one blog to another, one is likely to reach one of the top blogs within a few clicks of the mouse. Moreover, information on that top blog may have circulated into some of the blogs that linked to it, and from there propagated back to the blog where one started. Thus, even if one is not initially aware of a particular blog, one may end up there or be exposed to information posted on it. Overall, one is more likely to encounter a well-connected blog, or information posted on it, than one that is not.

The interactions with traditional media and social networks can further extend blog influence. For example, consider this sequence of events: A blogger posts an eyewitness account of events during Hurricane Katrina in 2005. A reader e-mails the story to associates, who forward it to others. One recipient is a television reporter, who reports the story on the evening news. Other journalists and bloggers pick up the broadcast. Soon it is all over the Internet, with commentary on some of the top blogs.

This behavior of jumping networks is common. It is illustrated by the CBS Memogate story in our introduction. In that case, the story jumped from a Web-based discussion group to blogs and traditional media.

Given the multiplicity of networks, their internal connectedness, and the ability of information to hop from one network to another, it would seem that information posted on a blog or elsewhere on the Internet could reach anywhere. In practice, however, an entity's influence – whether a blog, Website, newspaper, television station, person, and so on – is seldom so broad. In a study of six social networks, researchers found that individuals were unlikely to be aware of what others in the network were doing if they were more than two hops away. Thus, one's "sphere of observability" was restricted to direct connections and indirect connections involving at most one other person.[25]

The converse of this principle is that one's "sphere of influence" may likewise be limited to two hops on average. If so, bloggers whose posts are read by journalists working for broadcast media with large audiences can have a potentially much larger sphere of influence than those whose posts are read only by other bloggers or Internet users. We will return to this point in the discussion that follows.

Blogs vs. traditional media

There is a fierce debate among bloggers and traditional journalists about which media form is most important. Some bloggers, such as InstaPundit's Reynolds and other media watchers, cite a declining viewership of network news broadcasts[26] and a drop in circulation of daily US newspapers (overall circulation in 2005 was 55.2 million, down from 62.3 million in 1990[27]) as clear evidence that traditional media are on the way out, to be replaced by increasingly influential blogs and other Web-based media. Another measure is the numbers of bloggers who link to top news and media sites compared with those who link to the top blogs. Technorati found that the *New York Times*, CNN, and the *Washington Post* still garnered the most links, but the top blog (Boing Boing) beat out such media sites as PBS, Fox News, and *Time* magazine.[28]

The existence of a virtual universe of readily available, easily accessible sources of good-quality information may partially explain the dismal ratings traditional news media sources receive. Some Internet theorists (most notably, Howard Rheingold[29]) and market analysts have predicted that the plethora of choices for information consumers means that they will choose only those sources that confirm their own biases without challenging them; moreover, consumers will customize their newsfeeds so as to focus exclusively on a narrow band of interests. However, a recent Pew study[30] found that most Internet users do not consciously filter out opposing viewpoints and tend to encounter many diverse viewpoints through sheer accident. Furthermore, many news consumers in the Internet age want interaction, to be able to debate the stories and analyses that interest them. Interactivity is what blogs (and discussion groups) offer that video and print media do not.

On the other hand, there are good reasons to believe that reports of the traditional media's imminent demise are at best exaggerated, and at worst misleading. For example, the same industry data cited by blog proponents to show the decline in US newspapers circulation from 1960 to 2003 shows that in fact the decline occurred only for evening editions, whereas morning circulation actually grew and Sunday circulation remained generally constant.[31] Another reason is that despite the hype surrounding the emergence of blogs as a journalistic format, blogs have been clearly influential in only a few instances, while many other equally compelling stories never made the transition from blogs to the mainstream. This is an important point: Without the "mainstream" media's attention, Memogate and other stories that started on blogs might never have amounted to anything more than interesting but obscure historical footnotes.

W. L. Bennett's model of the infosphere provides a theoretical framework that explains why this is so. According to D. Travers Scott, Bennett stratifies the

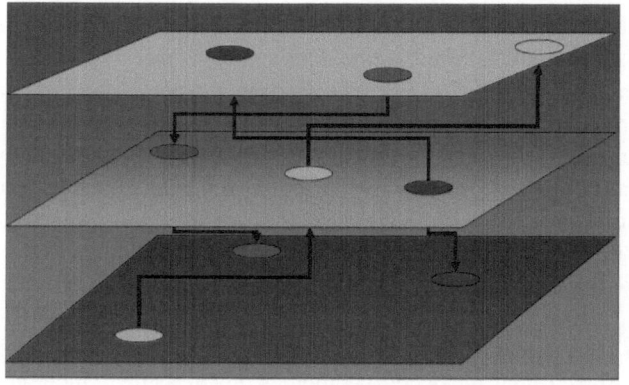

"Conventional Layer" (the Mediasphere): Mainstream, mass media

"Middle Layer" (the Blogosphere proper): Prominent blogs, webzines, advocacy groups

"Micro Layer" (other social uses of the Web): Email, mailing lists, personal blogs

Figure 9.3 W. L. Bennett's model of the infosphere (illustration created by the authors).

Infosphere into three layers (Figure 9.3): "the conventional layer of mainstream, mass media; the middle layer of prominent blogs, webzines, advocacy groups, etc.; and the micro layer of email, mailing lists, and personal blogs. The most successful communications strategies involve methods of getting a story to *access and activate all three strata*" (emphasis added).[32] Bennett's model is particularly useful because it offers a holistic view of the infosphere that incorporates everything from the mass media to the small-time blogger and taps into the social network to effect transitions between the layers. In this model, the mass media are vital to effective national political communication, consensus building, and mobilization.

Drezner and Farrell make the point that "under specific circumstances – when key weblogs focus on a new or neglected issue – blogs can socially construct an agenda or interpretive frame that acts as a focal point for mainstream media, shaping and constraining the larger political debate." They further note that a blog's influence has as much to do with the interactions between significant blogs and traditional media outlets as with the skewed distribution of links.[33]

Bloggers can also leverage the mainstream media by linking to selected news stories from their blogs. As a side effect, the stories' rankings by search engines such as Google may increase, making them more likely to come up in a Web search. Indeed, the *New York Times* reported in October 2006 that bloggers were attempting to influence the outcome of the November elections by boosting the Google rankings of articles that slammed certain candidates they opposed.[34] The goal was for users to be directed to the negative articles when they searched on the candidates' names.

Ultimately, each type of player has a vital role that benefits the system as a whole. The traditional media affect the *Zeitgeist* through print and broadcasting, whereas blogs and other Internet-based middle- and micro-layer media do the same online.

Measuring influence

If blogs can in fact be influential and may become more so in the future, as the anecdotal evidence suggests, it is important to find a way to measure a blog's influence as a determinant of the effectiveness of blogging as an IO weapon. In this section we describe various measures that are available today and others that may be worth developing. Our discussion does not catalog all commercially available products and services but attempts to cover the major metrics and some of their principal providers. We have divided them into two categories: measures based on site visits and measures based on links and citations.

Measures based on site visits include the following:

- *Visitors*. This is the number of different users who visit a blog over a given period of time. Alexa (www.alexa.com) computes this measure, which they call "reach," on a daily basis for Websites (not just blogs) visited by users of its toolbar.
- *Page views*. This is the number of pages viewed on a blog site over a given period of time. Alexa computes this measure on a daily basis for Web pages viewed by their toolbar users.
- *Traffic*. This is defined by Alexa as the mean of reach and page views averaged over a three-month interval and is the measure used to rank Websites.
- *Interactivity*. This is the degree to which a blog's readers offer feedback and participate in discussions with the blogger and other readers.

Alexa is the only source we found for metrics based on site visits. However, its metrics are not specific to blogs and are generally offered at the level of domain names rather than individual pages. Furthermore, they apply only to site visits by users of their toolbar. While the user base is fairly large (several million), it may not be representative of the total Internet population, which is about a billion.

We are not aware of any efforts to formalize a metric based on interactivity or to rank blogs accordingly. Nevertheless, this measure is potentially valuable, as high interactivity would suggest that visitors to a blog are sufficiently engaged to respond – and engagement in turn suggests influence.

Measures based on links and citations include the following:

- *Inbound blog links*. This is the number of other blogs with explicit links to a particular blog. Technorati ranks blogs on this basis, which it calls "authority." Blogstreet (www.blogstreet.com) uses inbound blog links to compute a blog influence quotient; however, it only counts links in blogrolls (lists of blogs), not links within blog posts.
- *Blog favorites*. This measures the popularity of a blog by its inclusion in a favorites list. Technorati has a favorites ranking based on information provided by their members.
- *Inbound Web links*. This is similar to inbound blog links but includes links

from anywhere on the Web: blogs, news sites, articles, personal pages, and so on. Google (www.google.com) computes this measure through its advanced search feature but does not offer a ranking of Websites based on the measure. PubSub (www.pubsub.com) has a similar measure, which it calls "InLinks." Commercial Web software tools such as Link Survey also compute this measure.

- *Weighted inbound Web links*. This is similar to inbound Web links but gives more weight to links from high-ranking sites. Google's PageRank metric falls in this category, as does PubSub's LinkRank. Blogrunner (www.blogrunner.com) is also said to have used a scheme of this type to identify influential bloggers. Their metric was calculated across posts over a sixty-day period.[35]
- *Search hits*. This is the number of hits obtained by searching on the blog name, with qualifiers as needed to distinguish it from other entities while eliminating hits within the blog's own site (the latter is easily specified with the advanced search option in Google). Search hits will pick up references to a blog that do not include explicit links, but it may also pick up references unrelated to the blog.
- *Media references*. This is the number of references to a given blog found in the non-Web media, including radio, television, magazines, journals, and newspapers.
- *Infosphere references*. This is the total number of inbound links and references from all sources in all three layers of the infosphere.

These measures each offer a different perspective on blog influence. For example, in April 2006, Technorati showed Boing Boing to be ranked number 1 in terms of authority (inbound blog links) but only third in terms of blog favorites. Google showed about 91,900 inbound Web links and a page rank of 9 (out of 10); it returned 10.5 million search hits for "boingboing -site:boingboing.net," which excludes internal references. On the other hand, Alexa reported a traffic rank of only 1,692, showing that whereas Boing Boing may indeed be popular among bloggers, it is viewed less frequently by Alexa users than nearly 1,700 other Websites.

We are not aware of any efforts to measure media references or infosphere references. However, these broader measures are important for determining influence. Looking back to Bennett's model, for a blog to be truly influential on a national scale, it has to reach and stimulate activity in all three layers of the infosphere, and it has to do this in a skewed system. New blogs constructed for the purpose of conducting PSYOP may have difficulty attracting enough links to reach all three layers. Moreover, the information content must stimulate interest and activity in all three layers, which it cannot do effectively if it appears to be propaganda. This limitation may be especially difficult to surmount because of the role that news bloggers tend to adopt: citizen journalists and media watchdogs.

Although most metrics are generic and not tailored to any specific topic, a corporate team comprising Market Sentinel, Onalytica, and Immediate Future has developed metrics for measuring the influence of bloggers with respect to a particular issue. Using keyword searches, they first build a list of stakeholder sites that have addressed the issue more than once and are cited by multiple sources. Then they examine three measures: numbers of citations by other stakeholders, how often a particular stakeholder serves as a broker of relevant information to another stakeholder (information influence), and the amount of influence given to a stakeholder by other important stakeholders (issue influence). The third metric appears to be related to the weighted inbound links metric described above. The metrics were used to show that Jeff Jarvis's blog, BuzzMachine, was the most influential source of information on the Internet regarding complaints about Dell Corporation's customer service in 2005.[36]

Being able to restrict the scope of a metric would be useful for influence operations. Besides topic area, metrics could be tailored to criteria such as nationality or demographics. For example, one might be interested in the most influential blogs written by Iraqi authors, read by Iraqis under age thirty, or receiving the most links from Arab bloggers. The most popular blogs overall might be irrelevant in certain regions or among certain audiences.

Another metric that could be useful, particularly when combined with those outlined above, is the frequency of postings on a particular topic or containing particular words or phrases. BlogPulse (www.blogpulse.com), for example, monitors the content of posts in the blogosphere in order to measure which topics are "hot." Users can also request a chart of the percentage of blog posts on a given topic plotted over a time interval of one to six months. A chart on the topic "Danish cartoon" created in early February 2006 showed a sharp upward spike in posts following eruption of the cartoon controversy. Being in a position to direct an already influential blog onto a hot topic could give an IO advantage in that domain.

In order to select or design a useful blog for conducting IO, it is critical to develop measures of the blog's ability to reach and activate the conventional, middle, and micro layers of the infosphere, as well as to develop a method for measuring the changes in attitudes, thinking, beliefs, and ultimately actions that are directly attributable to the blog. Selecting or creating a likely candidate blog is the first step, since the hard measure of effectiveness must necessarily come after an operation.

As noted earlier, a blog's popularity and credibility will be affected by the blogger's perceived reputation, knowledge, and skills. It is also likely to be affected by the perceived quality and credibility of the blog itself. Factors that could affect readers' assessments of a blog include:

- *Design*: the blog's general appearance and functionality, assessed according to the standards of the target audience. This includes such factors as quality of the graphics, professionalism, and ease of use.
- *Utility*: the relevance of the blog's content to issues of interest to the target community.

- *Accuracy and consistency*: whether information provided by the blog is generally correct or consistent with other information accepted by the target audience.
- *Currency*: the frequency with which the blog is updated.

Although we know of no studies that correlate these or other blog criteria with blog popularity or influence, we can draw on research conducted at Stanford on factors that affect the credibility of Websites. Researchers found that viewers paid particular attention to design but that utility and accuracy also mattered.[37] Postings that are well researched and thoughtful and that bear out this initial assessment over time may draw more links than ignorant diatribe, although this is not always the case.

Another obstacle to analyzing blogs by ranking is the recent emergence of splogs, or spam blogs, mentioned earlier. As David Kesmodel points out, some blog hosting services may lack sufficient controls to prevent automated software from creating splogs in bulk.[38] Because many page ranking systems are based on a site's number of inbound links, as described above, splogs can artificially inflate a site's Web status. Analysts must therefore ensure that all of the measures listed above, including the additional factors of design, utility, accuracy/consistency, and currency, are collected and correlated to avoid improper targeting for collection or exploitation.

Implications for information operations

We began with three research questions:

1 Are blogs truly influential, and if so, in what manner?
2 Does the information environment support blogging as part of an information campaign?
3 Can blogs provide a significant source of intelligence or measures of effectiveness for influence operations?

The answer to the first question is that a small number of blogs are, or have the potential to become, influential in the broadest sense. A blog's influence depends in large part on the reputation of its author and its content but also on the mathematical laws that govern the blogosphere as well as the blog's ability to reach and activate the three primary layers of the infosphere. Blogs that score high in the metrics identified above seem most likely to be those that have influence or become influential.

The answer to the second question is: It depends. One factor to consider is Internet penetration among the population of interest. In some parts of the world, only a small minority of the population has Internet access, and those who do may have similar wealth, education, experiences, and outlooks. In such cases, one cannot expect to reach the masses, and intelligence derived from blogs cannot be held to represent views of the majority of the population. On the other hand, as noted earlier, information posted on blogs can jump to traditional media and travel through social networks. Hence, even if members of a population do not have access to the Internet or read a particular blog, they may be influenced by it if local elites, including mainstream media elites, do.

Many countries censor the Internet, prohibiting their citizens from publishing certain information on a Website or blog or from accessing certain sites. In that environment, it may be hard to get a complete intelligence picture from bloggers inside the country or to influence the population with a blog that may be subject to censorship.

On the other hand, blogs that serve a small community or that occupy a specific niche may be useful for monitoring and targeting selected elements. People may serve in more than one social capacity; they may represent a class of community or peer opinion leaders and may be useful as both targets of influence operations and as vehicles for disseminating strategic communications.

A related point of difficulty in using blogs for IO is that segmentary opposition and its gentler cousin, in-group/out-group dynamics, may prevent a foreign audience from taking an overtly US government-run or sponsored blog seriously. Even American blogs show a high incidence of ethnic clustering,[39] and traditionally tribal societies, such as those of Afghanistan, are frequently defined by deep-seated fissures between major tribal groups. Even in settings where there are no widespread preconceptions about US use of propaganda, it may be easy for foreign audiences to dismiss the US perspective with, "Yes, but you aren't one of us, you don't really understand us."

Thus, in some settings, information strategists could consider clandestinely recruiting or hiring prominent bloggers or other persons of prominence already within the target nation, group, or community to pass the US message. In this way, the United States could overleap entrenched inequalities and make use of preexisting intellectual and social capital. Sometimes numbers can be effective; hiring a block of bloggers to verbally attack a specific person or promote a specific message, for example, may be worth considering.[40] On the other hand, such operations can have a blowback effect, as witnessed in the public reaction that followed revelations that the US military had paid journalists to publish stories in the Iraqi press under their own names.[41] People do not like to be deceived, and the price of being exposed is the loss of credibility and trust.

An alternative strategy is to "make" a blog and blogger. The process of boosting the blog to a position of influence could take some time, however, and could impose a significant educational burden, in terms of cultural and linguistic training, before the blog could be put online to any useful effect. Still, there are people in the military today who like to blog. In some cases, their talents might be redirected toward operating blogs as part of an information campaign. If a military blog offers valuable information that is not available from other sources, it could rise in ranking fairly rapidly.

Any blogs and bloggers serving an IO mission must be coordinated and synchronized with the overall influence effort in time and message. However, they must be prepared to argue and debate with their audience successfully and *inde-*

pendently on behalf of the US policy stance. In this sense, bloggers must be able to "circumvent the hierarchy," as blogger George Dafermos put it.[42] This means that they must be trusted implicitly to handle the arguments without being forced to communicate "solely by means of marketing pitches and press releases."

Beyond what might constitute "offensive" blogging initiatives, the United States needs also to be concerned with active defense measures. There are certain to be cases where a blog outside the control of the US government promotes or abets attacks against US persons or interests or actively supports the informational, recruiting, and logistic activities of our enemies. The initial reaction might be to seek to have the site shut down, but even successful efforts to do so do not guarantee that it will remain down. Many sites that have been forced to shut down have simply moved to a different host server, often in another country. Moreover, closing down a site would likely produce even more interest in the site and its contents. Also, taking down a site that is known to pass enemy EEIs (essential elements of information) and that gives us their key messages denies us a valuable information source. This is not to say that once the information passed becomes redundant or is superseded by a better source the site should be taken down. At that point the enemy blog might be used covertly as a vehicle for friendly information operations. Hacking the site and subtly changing the messages and data – merely a few words or phrases – may be sufficient to begin destroying the blogger's credibility with the audience. Better yet, if the blogger happens to be passing enemy communications and logistics data, the information content could be corrupted. If the messages are subtly tweaked and the data corrupted in the right way, the enemy may reason that the blogger in question has betrayed them and either take down the site (and the blogger) themselves or, by threatening such action, give the United States an opportunity to offer the blogger amnesty in exchange for information.

There also may be times when it is thought necessary, in the context of an integrated information campaign, to pass false or erroneous information through the media, on all three layers, in support of military deception activities. Given the watchdog function that many in the blogging community have assumed – not just in the United States, but around the world – doing so would jeopardize the entire US information effort. Credibility is the heart and soul of influence operations. In these cases, extra care must be taken to ensure plausible deniability and nonattribution and to employ a well thought-out deception operation that minimizes the risk of exposure. Because of the potential blowback effect, information strategy should avoid planting false information as much as possible.

This brings us to a more fundamental issue. Because the US military is prohibited from conducting information operations against US persons, it is reluctant to engage in Internet IO operations that might be characterized as PSYOP or deception. Once information is on the Internet, it can reach anyone, including people in the United States. Thus, while the military offers factual news on the Internet through public affairs programs, it generally stays away from commentary and IO. At least initially, this challenge might be addressed by using only accurate, factual information that is of some value to readers. Blogging can support public affairs efforts and focus on improving communications and building trust with local communities and the public. A blog can be used to solicit and respond to questions and concerns from target populations. In addition, military leaders might offer personal commentary on nonmilitary blogs, with the usual disclaimers.

Using blogs effectively in an information campaign may require a new intelligence tool, one that can monitor and rapidly assess the informational events occurring in a specific portion of the blogosphere and their effects, if any, on the three layers of the local infosphere.

Blogs and intelligence

Our third research question dealt with the suitability of blogs as a source of intelligence. We argue that blogs can provide excellent open-source intelligence, as well as serving as a nearly immediate measure of effectiveness for influence operations. Blog-derived intelligence can be considered a subset of both communications intelligence and open-source intelligence. We would expect to see it used primarily in support of information operations, although it does offer a broad range of possible applications. It may consist of computer network exploitation done in support of integrated PSYOP, public affairs, public diplomacy, military deception, and civil affairs/civil–military operations. The value of using blogs and blogging in support of a military information strategy depends heavily on the target region's Internet penetration and regulation (especially censorship). If the number of Internet users is small, it is necessary to determine who is using the Internet and for what purposes. Again, this very basic information should be collected as a part of the initial intelligence preparation of the environment, but once obtained it becomes a significant part of the baseline assessment for determining the need for, and the value of, conducting blog-based information operations.

If the initial assessment indicates that blogging activity is present, the next step is to look at the bloggers and their audiences, determine the blogs' functions (see Nardi et al.[43]), and conduct a preliminary analysis based on the metrics (blog visits, incoming links, and references) and indicators of quality and credibility (design, utility, accuracy, and currency) identified earlier. Questions that must be answered include the following:

- How large is the blogging community?
- Who are the bloggers? What are their positions and status within their communities and within the country as a whole (their general public roles and reputations)?
- Who is the target community or audience for each blog?
- Do the blogs address issues of social and political importance to the community they serve?
- What biases are observed in each blog? Do they reinforce or challenge the biases of their audience?

- Do any bloggers invite and engage in free and open interaction with their audience?

Answering these questions will require appropriate responses from intelligence agencies at all levels. National-level agencies are perhaps best suited to conduct comprehensive media, human factors, and social network analyses to identify and characterize the prominent or influential bloggers within their social networks and their connections to the larger community of traditional media journalists. These agencies should also examine the frequency with which each blog is referenced in the other media in the target region and perhaps engage in computer network exploitation to study reactions and references in the micro layer. Certain other existing assets, notably the Foreign Broadcast Information Service (FBIS) and the Armed Forces Information Service (AFIS), already collect thousands of broadcast and print news pieces on selected topics from around the world and could with relative ease look for blog references as well.

The importance of interactivity to a blog's influence is proportional to the size of the blog's potential audience. In areas where Internet access is limited to top government officials, interactivity may be of relatively low importance. In areas where a ruling or elite social class has access, interactivity can be more important. In countries where Internet access is widespread but free exchange of ideas is limited or discouraged, interactivity becomes golden. Again, the application of the general theory and principles must be flexible enough to account for differing political, social, and cultural conditions. Social network analysis and human factors analysis must be combined and correlated to craft specific messages to target specific bloggers and members of their audiences. Combined with a good sociological, psychological, and cultural framework for interpreting and predicting attitudes, behaviors, communications, and actions, intelligence derived from or about blogs can be highly effective in supporting influence and counterinfluence campaigns.

The entrenched inequality that characterizes the blogosphere has some implications for intelligence analysis and assessment. On the positive side, the fact that the most influential blogs generally have the highest number of links limits the number of blogs that must be read to glean the key or most widely held perspectives, concerns, attitudes, and knowledge that motivate the audience. A survey of only these blogs can serve as a rapid means of assessing the effectiveness of other influence operations in much the same way that a civil affairs soldier can assess general attitudes and mood by reading the graffiti on the walls. On the down side, in regions where Internet use is widespread, the tendency toward monopoly that results from systemic self-optimization will result in increasing homeostasis. Authors of top blogs in such environments may become disconnected from the content of their blogs and the concerns of their audience. This can occur when the demands of maintaining the blog override the blogger's ability to survey other blogs, conduct research, and maintain interactivity. Thus, in areas with widespread Internet use as well as in areas where only the elites or the government have Internet access, the content of the top blogs may not correlate well with majority concerns and opinions. In any setting, as always, intelligence drawn from blogs should be confirmed from other sources.

Analysts working with blog intelligence must have access to the operational disciplines they support; the closer the better. We recommend the creation of small special operations units with operational authority and integrated intelligence collections and analysis to conduct blog-based operations.

Recently, analysts at the Open Source Center under the auspices of the Director of National Intelligence have been monitoring and following significant foreign blogs and bloggers with the primary goal of exploiting them as sources of intelligence. A February 2006 posting on the organization's "Blog on Blogs" about Iranian expatriate blogger Hossein Derakhshan ("Hoder"), described him as "one of the most influential Iranian bloggers."[44] Open Source Center analysts used traffic data (based on reach and page views) from Alexa, authority rankings (inbound blog links) from Technorati, and frequency of postings from Blog-Pulse. All three sources offer the advantage of ready availability, but all have limitations, as discussed earlier.

Besides the various metrics that can be used, there are other tools that can help the intelligence analyst. BlogPulse's Conversation Tracker and TalkDigger (www.talkdigger.com), for example, track blog conversations as they spread through the blogosphere. TouchGraph (www.touchgraph.com) provides a tool for visualizing links among sites.

Blog-based operations

To function most effectively, units conducting blog-based operations must be staffed appropriately. Ideally, such units would be drawn from the special operations and intelligence communities, because of their historical experience in sensitive operations. Linguists and intelligence analysts – ideally, analysts who are also linguists – who are commanded or advised by a qualified PSYOP or IO officer should form the core of such a unit. The unit's capabilities must be augmented through liaison relationships with the other influence organizations responsible for planning and conducting PSYOP, public affairs, public diplomacy, civil affairs/civil–military operations, and military support to public diplomacy. Because of the unique nature of intelligence support to blog-based operations, comprising both open and highly classified sources, and the necessity of producing a product intended for open distribution, a blog operations unit should have solid information, operations, and network security programs in place. Just as other US government intelligence or special activities require oversight by the House and Senate Intelligence Committees to prevent potential abuses, and just as US military activities are subject to review by a judge advocate and an inspector general to ensure compliance with laws and regulations, so too should blog-based operations units require oversight and regulation.

In order to act and react efficiently in managing bloggers and blogs, the intelligence specialists and planners who have the knowledge should run the actual blog. In cases where indigenous bloggers and their blogs have been identified and recruited, the blog operations cell should also house the case officer managing the asset. The same metrics used to select a blog can also serve as indirect measures of effectiveness. For example:

- Once blog operations have begun, does the blog attract new inbound links?
- Is there an increase over time in the blog's ranking via various metrics?
- Through polling and media analysis, can a change in public opinion be correlated with growth in the blog's indicators?
- What does content analysis of the interaction that occurs with the blogger on the site reveal? (Change in opinions posted by readers? Positive or negative?)
- Do the comments on the blog correlate with public opinion results obtained by polling or portrayed in the mainstream media?
- Is the blog referenced by the mainstream media in the target country? If so, how often?
- Do other sources of intelligence confirm these indicators?

Like any other influence operation, blog operations must be given time to work. There is no magic formula. We would suggest conducting quarterly reviews of the blog's effectiveness along these lines and then making adjustments to reverse any negative trends and accelerate positive trends.

This fusion of intelligence and operations is a requirement for operating in a medium that rewards the efficient distribution of knowledge and information above all other considerations, and it is in the best traditions of the intelligence and special operations communities. Pushing operational authority out to those best equipped to receive, analyze, and act in a dynamic information environment maximizes both efficiency and effectiveness. Although a blog-based operations unit could be based either domestically or in theater, the best option is to forward deploy it as a cell, just as we deploy our PSYOP analysts and production and dissemination capabilities.

Conclusion

In this chapter, we outlined some of the potential uses of the blogging phenomenon for military information strategy. While we can make reasonably certain statements about blogging in general that can be applied universally, the actual employment of tactics and techniques comes down to cases. The successful employment of blog operations depends on a number of variables that must be determined during intelligence preparation. In environments that support them, properly conducted blog operations have the potential to be extraordinarily powerful, as demonstrated by the high-profile incidents involving blogs during the 2004 presidential campaign.

Some of the possible techniques we explored in our discussion of the military use of blogging require a certain degree of subtlety, finesse, and covert action. By giving military blog-based operations to the Intelligence and Special

Operations communities, such operations become less risky and more feasible. However, military operations must necessarily remain only a part of a larger effort. Given US and international law and the distribution of the necessary authorities among many (often competing) government agencies, any conduct of influence operations through the blogosphere will require a truly integrated interagency approach and thus belongs at the national level as a part of an over-arching strategic communications effort.

We also identified several areas that may be of interest to other researchers. Chief among these are the questions of how various blog metrics relate to each other and to actual influence as determined by changes of behavior in a target population. Also, although we identified what we think are significant indicators of a blog's influence, techniques for measuring them and testing our hypothesis largely remain to be explored. Finally, the suggested techniques for using computer network attacks against blogs in support of influence operations are hypothetical as well and require testing before they can be employed.

One of the significant limitations of our presentation as an initial foray into military use of the blogosphere is that much of the information available concerns American blogs, run by Americans, largely for an American audience, whereas military use of the blogosphere must necessarily focus on foreign blogs, bloggers, and audiences. However, because some factors, such as the scale-free nature of the Internet and the psychological basis of influence, are universal, we hope with this chapter to lay a general foundation for military use of the blogosphere that can be adapted to specific tactical circumstances by information operators.

Notes

1 "The Sixty-First Minute," Power Line, 9 September 2004, powerlineblog.com/archives/007760.php (accessed 19 April 2006).
2 K. Parker, "Bloggers Knew!" TownHall.com, 15 September 2004, www.townhall.com/columnists/KathleenParker/2004/09/15/bloggers_knew! (accessed 20 September 2005).
3 C. Pein, "Blog-Gate: Yes, CBS Screwed Up Badly in 'Memogate' – But So Did Those Who Covered the Affair," *Columbia Journalism Review Online*, January–February 2005, www.cjr.org/issues/2005/1/pein-blog.asp (accessed 20 September 2005).
4 T. Thornburgh and L. Boccardi, "Report of the Independent Review Panel on the September 8, 2004, *60 Minutes Wednesday* Segment 'For the Record' Concerning President Bush's Texas Air National Guard Service," 5 January 2005, www.image.cbsnews.com/htdocs/pdf/complete_report/CBS_Report.pdf (accessed 19 September 2005).
5 S. Kenny, "The Revolution Will Be Blogged," *Salon*, 6 March 2006, www.salon.com/news/feature/2006/03/06/iranian_bloggers/index_np.html (accessed 18 April 2006).
6 L. Shane, "Military Issues Content Warning to Combat-Zone Bloggers," *Stars and Stripes*, 4 October 2005, www.estripes.com/article.asp?section=104&article=31111&archive=true (accessed 6 October 2005).
7 A. L. Barabási and E. Bonabeau, "Scale-Free Networks," *Scientific American*, May 2003, pp. 60–9.

8 D. Sifry, "State of the Blogosphere, February 2006, Part 2: Beyond Search," Technorati Weblog, 13 February 2006, www.sifry.com/alerts/archives/000420.html (accessed 8 May 2006).

9 D. Drezner and H. Farrell, "The Power and Politics of Blogs," presented at the 100th annual meeting of the American Political Science Association, Chicago, 2–5 September 2004, www.utsc.utoronto.ca/~farrell/blogpaperfinal.pdf (accessed 2 October 2005).

10 D. Sifry, "State of the Blogosphere, February 2006, Part 2: Beyond Search," 14 February 2006, www.technorati.com/weblog/2006/02/83.html.

11 Barabási and Bonabeau, "Scale-Free Networks."

12 Drezner and Farrell, "The Power and Politics of Blogs."

13 Sifry, "State of the Blogosphere, February 2006."

14 Pew Internet and American Life Project, "The State of Blogging," January 2005, www.pewinternet.org/pdfs/PIP_blogging_data.pdf (accessed 19 Apr 2006).

15 D. Sifry, "State of the Blogosphere, April 2006, Part 1: On Blogosphere Growth," Technorati Weblog, 17 April 2006, technorati.com/weblog/2006/04/96.html (accessed 18 April 2006).

16 D. Kesmodel, "'Splogs' Roil Web, and Some Blame Google," *Wall Street Journal Online*, 19 October 2005, online.wsj.com/public/article/ SB112968552226872712–8b5l_fijhNltE4s7DX6tvLI9XNo_20061025.html (accessed 26 October 2005).

17 Sifry, "State of the Blogosphere, April 2006."

18 Pew Internet and American Life Project, "A Decade of Adoption: How The Internet Has Woven Itself into American Life," 25 January 2005, www.pewinternet. org/PPF/r/148/report_display.asp (accessed 21 September 2005); H. Lebo and S. Wolpert, "First Release of Findings from the UCLA World Internet Project Shows Significant 'Digital Gender Gap' in Many Countries," *UCLA News*, 14 January 2004, www.uclanews.ucla.edu (accessed 14 January 2004).

19 Pew Internet and American Life Project, "The State of Blogging."

20 B. Nardi, D. Schiano, M. Gumbrecht, and L. Swartz, "I'm Blogging This: A Closer Look at Why People Blog," 2004, www.ics.uci.edu/~jpd/classes/ics234cw04/ nardi.pdf (accessed 15 September 2005).

21 S. Turkle, *Life on the Screen: Identity in the Age of the Internet*, New York: Simon & Schuster, 1995.

22 K. Gill, "How Can We Measure the Influence of the Blogosphere?" May 2004, p. 8, faculty.washington.edu/kegill/pub/www2004_blogosphere_gill.pdf (accessed 29 September 2005).

23 "Jeff Gannon," *Wikipedia*, en.wikipedia.org/wiki/Jeff_Gannon (accessed 5 May 2006).

24 Nardi *et al.*, "I'm Blogging This."

25 N. E. Friedkin, "Horizons of Observability and Limits of Informal Control in Organizations," *Social Forces*, 62:1 (September 1983): 54–77.

26 L. Lazaroff, "Audience Decline an Old Story," *Chicago Tribune Online*, 24 November 2004, www.csupomona.edu/~jrballinger/448/readings/News_Audience_ decline.htm (accessed 15 August 2005).

27 G. F. Will, "Unread and Unsubscribing." *Washington Post*, 24 April 2005, www.washingtonpost.com/wp-dyn/articles/A10698–2005Apr22.html (accessed 15 August 2005).

28 Sifry, "State of the Blogosphere, February 2006."

29 H. Rheingold, *Smart Mobs: The Next Social Revolution*, New York: Perseus Publishing, 2002.

30 Pew Internet and American Life Project, "The Internet and Campaign 2004," 6 March 2005, www.pewinternet.org/pdfs/PIP_2004_Campaign.pdf (accessed 8 May 2006).

31 Newspaper Association of America, "Facts About Newspapers," 2004, www.naa.org/info/facts04/circulation-daily.html (accessed 8 May 2006).

32 D. Scott, "Tempests of the Blogosphere: Presidential Campaign Stories That Failed to Ignite Mainstream Media," MIT, 2005, web.mit.edu/comm-forum/mit4/papers/scott.pdf (accessed 20 September 2005).

33 Drezner and Farrell, "The Power and Politics of Blogs."

34 T. Zeller Jr, "A New Campaign Tactic: Manipulating Google Data," *New York Times*, 26 October 2006.

35 Gill, "How Can We Measure the Influence of the Blogosphere?"

36 Market Sentinel, Onalytica, and Immediate Future, "Measuring the Influence of Bloggers on Corporate Reputation," December 2005, www.marketsentinel.com/files/MeasuringBloggerInfluence61205.pdf (accessed 8 May 2006).

37 B. J. Fogg, C. Soohoo, D. Danielson, L. Marable, J. Stanford, and E.R. Tauber, "How Do People Evaluate a Web Site's Credibility?" Consumer Reports WebWatch, 29 October 2002, www.consumerwebwatch.org/dynamic/web-credibility-reports-evaluate-abstract.cfm (accessed 5 October 2005).

38 Kesmodel, "'Splogs' Roil Web, and Some Blame Google."

39 H. Choi, "Ethnic Clustering in Blogging Communities," 2003, www.students. haverford.edu/hchoi/final%20project.htm (accessed 30 September 2005).

40 D. Lyons, "Attack of the Blogs," Forbes Online, November 2005, www.forbes.com/forbes/2005/1114/128_print.html (accessed 5 November 2005).

41 M. Mazzetti and B. Daragahi, "U.S. Military Pays to Run Stories in Iraqi Press," *Los Angeles Times*, 30 November 2005.

42 G. Dafermos, "Blogging the Market: How Weblogs Are Turning Corporate Machines into Real Conversations," 2004, opensource.mit.edu/papers/dafermos3.pdf (accessed 28 September 2005).

43 Nardi *et al.*, "I'm Blogging This."

44 B. Khanoon, "Hoder: One of the Most Influential Iranian Bloggers," Blog on Blogs, DNI Open Source Center, Washington, DC, 12 December 2005, https://www.opensource.gov (accessed 9 May 2006.)

Conclusion
Why is information strategy difficult?

Douglas A. Borer

Strategy is a weighty word that carries significant influence in any discourse. If something is termed "strategic," our society and government deem it to be of essential importance. In the realm of international politics, things strategic are most often related to grim matters of life and death, peace and war. Thus, "information strategy" is also of a matter of significance, and one that policy makers and policy executors would be expected to place heavy emphasis on mastering. Regrettably, as many of the contributors to this volume suggest, the United States seems to have particular difficulty in devising and implementing an information strategy. Why is this so?

Why strategy is difficult

To begin with, strategy itself is a difficult endeavor.[1] Moreover, the term itself has been defined in numerous ways, linking various concepts together in manifold ways. Practitioners of military operations may favor Clausewitz's view that "strategy is the use of engagements for the object of war." For students of international relations, Gregory Foster's description, "Strategy is ultimately about effectively exercising power," might suffice, unless Bernard Brodie's "strategic theory is a theory of action" is preferred. Military planners, physicists, economists, business leaders, mathematicians, historians, political scientists, and members of any number of other professions have contributed to the rich body of literature on strategy.[2] What this literature shows is that while many people readily agree that thinking strategically is a necessity, few agree on what strategy means and how it is to be conceptualized, executed, and evaluated after implementation.

One simplified approach, taught for years by Col. (Ret.) Arthur Lykke at the US Army War College, purports that strategy of any type consists of objectives or ends, concepts or ways, and resources or means.[3] This tripartite framework illustrates how strategy is fundamentally a calculated relationship between ends and means. The practitioners of strategy ask, "What do I want, and how am I going to get there?" Lykke's ends-ways-means approach is valuable for its insistence that one think in a fairly rigorous and compartmentalized manner about the process of strategizing. For instance, it is clear that President George W. Bush's major objective after the 11 September 2001 terrorist attacks was to protect the United States from further attacks by would-be terrorists. This worthy goal was the initial step of strategizing. How was he to achieve this end? To answer that question, the president and his trusted advisers first had to consider concepts and ways, beginning by answering the question, "What causes terrorism?" In relatively short order, and with minimal input from academic specialists with culturally specific knowledge of the Middle East and Islam, they settled on the idea that Islamic terrorism springs primarily from the frustration of individuals who are not satisfied with their own governments. Their frustration in turn is driven primarily by a deficit of political participation; had an adequate level of political participation existed, it would have preempted violence by channeling dissatisfaction, anger, and frustration through more effective – and more peaceful – social and governmental institutions and processes.

Indeed, the Bush administration believed that "bad" governments, meaning nonrepresentative ones (old-style monarchies, dictatorships, single-party authoritarian states, and so on) are ultimately to blame for terrorism.[4] Starting from this premise, the administration then moved further into step two of Lykke's framework, in which ways of dealing with the problem were considered. Here, no great leap of intellect was needed: if "bad" governments were to blame, the logical solution was for them to be replaced by "good" governments. From an American perspective, the only genuinely "good" government structure is one that exhibits free, fair, and openly contested elections, one in which the rule of law guarantees basic civil and human rights (as understood in the West), and one in which the role of government is primarily to ensure that the individual is protected, nurtured, and emancipated from undue governmental, social, or ideological constraints (and the individual could in turn be expected to respect the rights of others). In short, "good governments" are those that are founded on a genuinely robust "social contract" of the sort found in Western liberal democracies. Thus, the spread of American-style democracy throughout the Middle East became the major way that future terrorist strikes against the United States could be averted.[5]

What then of the final step of Lykke's framework: organizing the resources, or the means, to execute the ways that had been chosen for achieving the ends? Generally, this category is thought of today in terms of the elements of national power, a concept that observers of international politics have inconclusively debated since ancient scribes first copied and distributed Thucydides' *History of the Peloponnesian War* some twenty-five centuries ago. In the United States, military officers and diplomats are trained to think of power in terms of the "DIME" acronym, in which diplomatic, informational, military, and economic tools of statecraft are to be considered when attempting to influence other actors in the international system. For the Bush administration, the military component was clearly perceived not only as the correct tool but, as is now apparent, the *only* truly necessary tool for achieving American national security after 9/11. As a result, a "global war on terrorism" was declared, Afghanistan and Iraq were invaded, and US military and intelligence agencies began vigorously to conduct global operations "in the shadows" throughout various locales where terrorists were thought to lurk.

With the advantage of hindsight, it is now clear (some six years later) that the American strategy was deeply flawed. Certainly Bush's aspiration to protect Americans from further 9/11's was a perfectly worthy and moral goal, and, moreover, one that the president was constitutionally duty-bound to pursue. However, in terms of strategic performance, it is abundantly clear that the ways and means of attaining it were ill chosen.

Leaving aside debates over the in's and out's of democratic theory, how democracy is spread, and what virtue democracy may or may not have in transforming societies and individuals, there is no denying that since the United States embarked on the present strategy, democracy has broadened in the Middle East (to Afghanistan, to Iraq, and to the Palestinian territories), but terrorism and rebellion have also increased, contrary to the promised results of the president's strategy.[6] Given this outcome, it does not seem amiss to suggest that US national interests may be better served by dealing with current "bad" regimes in the Middle East (America's traditional allies in Egypt, Saudi Arabia, Jordan, Morocco, and so on) than with potential democratic ones in these countries. Pakistan and Algeria are both examples of the United States aligning itself with secular military governments that have acted forcefully against the democratic process. As Gregory Gause aptly observed,

> The problem with promoting democracy in the Arab world is not that Arabs do not like democracy; it is that Washington probably would not like the governments Arab democracy would produce.... Further democratization in the Middle East would, for the foreseeable future, most likely generate Islamist governments less inclined to cooperate with the United States on important US policy goals, including military basing rights in the region, peace with Israel, and the war on terrorism.[7]

The lesson for Washington's planners is clear: "Be careful what you wish for."

The recent democratic rise to power of Hamas in Palestine provides an illustrative case: in electing Hamas, the Palestinian people chose terrorists as their leaders. Moreover, newly democratically elected regimes (like Hamas, and like Hizbullah in Lebanon) are clearly capable of supporting and sponsoring terrorism. In other words, not only might newly elected democracies not be US allies, they may actually become enemies. Islamic extremist parties may come to power and, as the Iranian revolutionaries did in Tehran after 1979, change the foreign policies of their countries and fund, support, and use terrorists as proxies for their political battles with other countries, including the United States. Democracy may in fact be a "way" of spreading rather than reducing terrorism.

The error of choosing the forcible spread of democracy as the way to fight terrorism was compounded by the decision to rhetorically declare a "global war on terrorism." Declaring war on a form of warfare, rather than on an identifiable enemy that might be defeated, is problematic. Indeed, in the terminology of warfare, terrorism is a tactic, and declaring war on it is equivalent to declaring war on the flanking maneuver, on precision bombing, or on the surprise attack. Such declarations may sound good for a short time to an anxiety-ridden and revenge-prone public, but they make no sense as a guide to strategy. However, as suggested above, the major error here was the choice to emphasize military force rather than the diplomatic, economic, and (most importantly in the collective opinion of the contributors to this volume), informational aspects of national power to combat terrorism. As the old saying goes, it is a bad idea to bring a knife to a gun fight; so too it is a bad idea to use a conventionally armed, equipped, trained, and organized military force as the principle means to fight what is primarily an information war – a war that requires a more nuanced, sophisticated, flexible, and distributed approach.

Why information strategy is difficult

This is not to say that the US military has no role in combating terrorism – far from it. However, in an information war, the US military's doctrinaire preference to "find, fix, and finish" the enemy is rarely applicable, and in most cases it is counterproductive. If information is the key to combating terrorism effectively, why does the United States have so much difficulty with the concept? But let us first back up a step and ask, "What exactly do we mean by information"?

Clearly, no single answer will satisfy everyone. What is information today will still be information tomorrow, but the *effect* of a given piece of information may be very different from one day to the next, depending on the time, the circumstances, the actors involved, and, most important, who the consumers or recipients of that information are. For instance, on a given day a news story may report that a pair of hijacked airplanes crashed into the twin towers of two of the tallest buildings in the world, killing all on board and 3,000 or so in the buildings. The planes were reportedly hijacked and piloted by radical Islamists based in Afghanistan who were bent on influencing the behavior of the government of the state on whose territory the buildings stood. This scenario describes the events of 9/11, and we know what the subsequent course of events has been. However, consider what would have happened if all parts of this news story were the same as the 9/11 version except that the location of the attack was the Petronas Towers in Kuala Lumpur, Malaysia, rather than the World Trade Center in New York City. Certainly the course of history would have been affected if terrorists had attacked the secular government of Malaysia with the intent of bringing it down, but by no stretch of the imagination would world history have unfolded as it has over the past six years.

The point is that information, like power, must be understood as a *relational* concept rather than an absolute one. Just as with power, breaking this concept down into its component parts is helpful to allow strategists to make better choices. But information, like power, is intrinsically difficult to measure, categorize, and understand, and it is very hard to use effectively in a predictive or linear fashion. What gives you power today might leave you powerless tomorrow. What constitutes good information today may, depending on the audience, be misinformation next week. The contributing authors to this volume have deliberated on this relational problem from a variety of perspectives that collectively illustrate the need for a broader understanding of information itself. I agree with Barrett and Sarbin's observation in chapter 1 that because ideas are fundamentally persuasive rather than coercive, the notion of a "war of ideas" may be an oxymoron, or at least an inherently misleading approach to the role of information. However, information writ large – in which ideas are a subcomponent – is an integral and defining part of warfare; in this broader perspective, it certainly makes sense to give ample attention to the "information war."

Incendiary bombs dropped on London by German airplanes during the blitz were bits of information, not only in terms of the story they generated in the newspapers but also as a psychological message being sent from the German government to the British government and people. Likewise, the "shock and awe" campaign, for which the Pentagon trumpeted victory even while the fighting was still taking place, was a package of information meant to influence the Iraqi government and people. Yet both these examples suggest that the recipients of information often draw a very different message from it than that intended by the producer. The British did not sue for peace in 1940, and the Iraqis were neither shocked nor particularly awed in 2003, and after an initial period of confusion, their anger grew into an effective insurgency. It may be axiomatic that harnessing the power of information in terms of desired outcomes is much more of an abstract art than it is a predictive science.

The battle for the story

It is clear that in an era in which traditional nation-states are increasingly pitted against loosely affiliated terrorist networks at the local, regional, and global levels, terrorist organizations currently have an information advantage over states. As argued throughout this book, the United States and its allies can and should make efforts to reverse this trend, first by becoming less hierarchical and more networked themselves – an approach that would enable them to leverage the significant technological resources they possess. However, states being what they are, such change would occur only haltingly at best in the face of the natural tendencies states have toward hierarchy and centralization. Thus, in terms of information warfare, the comparative advantage of states may in fact be first to rejuvenate older, more idea-based concepts, such as deception, psychological operations, and a renewed commitment to Cold War-style public diplomacy. The chapters in this volume by Whaley, Rothstein, and Lord suggest that the basic tool kit for conducting information warfare has been emptied of most everything but the hammer. Sadly, as Lord recounts so well, the United States unilaterally disarmed itself in the area of public diplomacy (a Cold War strength) at the very same time that its enemies launched a major arms acquisition campaign. Until America's informational disarmament is reversed, its opponents will have the operational advantage.

In addition, no matter how well the collective components of the US government that represent the DIME tools of statecraft are organized and resourced, what we are really talking about here, what the essence of information strategy is fundamentally, comes down to the *story* we are trying to tell the world. More importantly, whatever the story we are telling, to what extent do our actions validate the veracity of our words? One of the unforeseen results of globalization in the "information age" is that historical double standards (of which all states have been guilty of throughout the ages) are increasingly difficult to hide or explain way. As states lose control over information and individuals gain ever more control, increasing numbers of the world's citizens have access to good (and bad) information about the world around them. During the Cold War, the United States was able to combat communism effectively, even though it meant aligning itself with right-wing authoritarian dictatorships that were every bit as vicious as Moscow and Peking's left-wing allies. However, Washington's inconsistencies were always seen as the lesser of evils when communism was the alternative. With the end of the Cold War, America's Cold War-era double standards have come under intense scrutiny by the consumers of information in the twenty-first century, including the 1.3 billion Muslims in the world.

Indeed, a critical flaw in America's information strategy today is derived in large part from the gross injustice that many Muslims perceive in US policy in the Middle East. Despite the wide array of religious opinions among Islam's various sects and subsects, there is a single viewpoint that is shared by nearly all. Muslim Jordanians, Muslim Malaysians, Muslim Montanans, Turkish Muslims, Trinidadian Muslims, Tanzanian Muslims, and Muslims everywhere agree that the Palestinians are getting a raw deal from the Israelis and that the US government's words simply do not reflect its actions when it comes to judging all countries equally by America's own self-professed standards. Even though the United States promotes democratization in the Middle East and has invaded two Muslim countries to bestow the blessings of democracy on their people, Washington rejected the outcome of the democratic process in Hamas's parliamentary victory in 2006. Washington also fails to recognize that Iran is a democracy (albeit one of a different sort). Moreover, in Afghanistan and Iraq the democracy that has been imposed by force appears to be a contributing cause of the ongoing Islamic insurgencies that threaten to spread throughout the region. Furthermore, as noted above, the United States continues to lend strong support to traditional authoritarian governments throughout the Muslim world, with Washington seeing any pro-US government as the genuine definition of a "good government," no matter how antidemocratic it may be. Many Muslims around the world are thus justifiably cynical when they hear American leaders exhort them to become more democratic. Most Muslims abhor the extreme violence of the jihadists, but they are also repulsed by the injustices they perceive as being perpetuated by Israel against the Palestinians as well as by their own governments against domestic political dissent. Until the United States convincingly gets this story straight, the information war against the jihadists will be most difficult to win among rank-and-file Muslims.

Globalization is largely driven forward by the influence that new forms of information technology have had on economic, social, and political systems. Among other things, what we are witnessing in the Middle East and beyond today is a primitive, highly politicized faith-based reaction among Islamic radicals to globalization itself. Liberal capitalism has now pushed into every corner of the globe, bringing goods, services, ideas, images, and behaviors into the homes of people of all cultures. Indeed, Western proponents of globalization have often argued that such a transformation would bring people together by showing them that all human beings are essentially the same. However, for people like Osama bin Laden and his followers, the opposite is true: globalization has forced upon the Muslim heartland a degenerate secular godlessness. Globalization has brought us closer together, and we have discovered that we are different. Indeed, for jihadists, Western liberal culture is seen as an insidiously seductive brew of vice, decadence, and greed that offends God and corrupts the faithful.[8] It is to be fought by all means available. Indeed, as Ronfeldt suggests in Chapter 2, it seems that globalization has bred a new tribalism.

The perverse irony of today's information war is evident. Jihadist networks are spawned and populated by people who are fiercely opposed to globalization and who want to establish neotraditional forms of rule that empower religious rather than secular authority. However, these jihadist networks could not survive, thrive, and prosper without the Internet, modern communications, and the free flow of people, ideas, and capital that characterize globalization in the information age. Whether or not the jihadists will be successful in their quest to turn back the march of civilization, with religion again becoming a dominant force in world politics, is anyone's bet. However, what is clear is that inside Islam, the Muslim World is entering a new phase of internal turmoil between Sunni and Shi'i sects that may turn out to be every bit as destructive as the Thirty Years' War (1618–48) in Europe. Although this war was primarily a religious conflict between Catholics and Protestants, it also involved balance-of-power intrigues that witnessed Catholic France siding with German Protestants against the Holy Roman Empire. The mixture of religious zeal and great-power international politics eventually cost Europe as much as a third of its population.[9]

Will the Sunni–Shi'i conflict now taking root in Iraq evolve into a broader intra-Islamic war that could have as much an impact on the world as the Thirty Years' War did? For the United States, the first challenge may simply be to realize that despite 9/11, and despite the nation's vast resources as the global hegemon, this fight is now increasingly about issues and between populations over which Washington, London, and the other power centers of the West have minimal influence. Indeed, the great challenge of information strategists today may be the demand for its practitioners to develop a hitherto unimagined capacity for patience, restraint, and nonkinetic responses to terrorist violence. Sun Tzu perhaps said it best centuries ago in *The Art of War*: "He who knows when he can fight, and when he cannot, will be victorious."

Notes

1 See David Jablonsky, "Why Is Strategy Difficult?" In *US Army War College Guide to Strategy*, Carlisle, PA: Strategic Studies Institute, 2001, pp. 143–56.
2 See Colin S. Gray, *Modern Strategy*, New York: Oxford University Press, 1999, pp. 17–20.
3 See H. Richard Yarger, "Towards a Theory of Strategy: Art Lykke and the Army War College Strategy Model." Available at dde.carlisle.army.mil/authors/ stratpap.htm.
4 *The National Security Strategy of the United States*, September 2002. Available at www.whitehouse.gov/nsc/nss.pdf.
5 Ibid.
6 See Michael Freeman and Douglas A. Borer, "Thinking Strategically: Can Democracy Defeat Terrorism?" In James J. F. Forest, ed., *Countering Terrorism and Insurgency in the 21st Century*, vol. 1, *Strategic and Tactical Considerations*, New York: Praeger, 2007, chap. 3.
7 F. Gregory Gause III, "Can Democracy Stop Terrorism?" *Foreign Affairs*, September/ October 2005.
8 See Douglas A. Borer and Mark T. Berger, "All Roads Lead to and From Iraq: The Long War and the Transformation of the Nation State," Third World Quarterly, vol. 28, no. 2, 2007.
9 See Paul Kennedy, *Rise and Fall of the Great Powers*, New York: Random House, 1987, pp. 36–72.

Index